ABSTRACTS OF PAPERS PRESENTED IN SYMPOSIA

XVIIth International Congress of History of Science
University of California, Berkeley
31 July - 8 August 1985

Acts, Volume 2

Office for History of Science and Technology
University of California, Berkeley
1985

Copyright © 1985 by the Regents of the
University of California

ISBN # 0-918102-14-6

Library of Congress catalog
card no. # 85-62072

PREFACE

This volume contains abstracts of papers presented to the Symposia of the XVIIth International Congress of History of Science. All abstracts received by 15 June 1985 are included.

The abstracts are arranged by Symposium, and within the Symposium by session. An alphabetical index by author at the end of the volume gives the number of the Symposium and the session in which the author's paper is scheduled. The Symposium and session numbers, in the form "17.2", are also entered at the foot of each abstract to serve as pagination.

CONTENTS

No. 1 SCIENCE AND TECHNOLOGY IN THE MIDDLE AGES
    Session 1 Alfonso X and His Era
    Session 2 Mechanical Arts in the Middle Ages and the Early Renaissance

No. 2 TECHNOLOGICAL TRAINING AND EDUCATION: NATIONAL COMPARISONS
    Session 1 Technological Training and Industrial Development
    Session 2 Engineering Education in National Contexts

No. 3 GENETICS AND SOCIETY
    Session 1 Genetics, Biotechnology, and Public Policy
    Session 2 Genetics and Cytology
    Session 3 Eugenics
    Session 4 Human and Medical Genetics
    Session 5 Molecular Genetics and Genetic Engineering
    Session 6 Genetics, Horticulture, and Agriculture

No. 4 WOMEN IN SCIENCE: OPTIONS AND ACCESS
    Session 1 History of Science and Mathematics Education for Women and Girls: Sex-Typed Textbooks and Curricula
    Session 2 History of Science and Mathematics Education for Women and Girls: Foreign Study--Degrees and Leverage
    Session 3 Women and Technology in Early Modern Times
    Session 4 Women and Technology: Spheres of Creativity
    Session 5 Fields and Specialities In Which Women Were Numerous Successful: Mathematics and Physical Sciences
    Session 6 Fields and Specialities In Which Women Were Numerous Successful: National Trends

No. 5 GOVERNMENT, INDUSTRY, AND THE GROWTH OF COOPERATIVE RESEARCH
 Session 1 Military Research
 Session 2 Space Research
 Session 3 National Laboratories
 Session 4 Organized Research and Policy

No. 6 CROSS-CULTURAL TRANSMISSION OF NATURAL KNOWLEDGE AND ITS SOCIAL IMPLICATIONS
 Session 1 India
 Session 2a East Asia
 Session 2b East Asia
 Session 3a Latin America: General Considerations
 Session 3b Latin America: Reception of European Science
 Session 4 Islam

No. 7 UNDERSTANDING AND USES OF NATURE IN NATIVE CULTURES
 Session 1 Perceptions of Earth and Sky
 Session 2 Interaction of Human and Natural Environments

No. 8 WESTERN SCIENCE IN THE PACIFIC BASIN
 Session 1 Exploration and Navigation
 Session 2 Ethnology, Natural History, and Oceanology
 Session 3 Organization of Scientific Research in the Pacific Basin

No. 9 EARTH SCIENCES IN THE 19th AND 20th CENTURIES
 Session 1 Petroleum Exploration and Seismology
 Session 2 Education in Geology

No. 10 SCIENCE, LITERATURE, AND THE IMAGINATION
 Session 1 Science Fiction
 Session 2 Literature and Science
 Session 3 The Sciences and the Arts
 Session 4 Movies, Poetry Reading, and Dedication of the New Society

No. 11 HISTORICAL SOCIOLOGY OF SCIENCE
 Session 1 Ideologies and Scientific Development
 Session 2 The Laboratory and the Construction of Knowledge

No. 12 SCIENTIFIC INSTRUMENTS
 Session 1 Development of Instruments in Cultural Contexts
 Session 2 Scientific Instrument and Research Methods
 Session 3 Instruments and Artifacts as Historical Evidence

No. 13 PUBLICATIONS
- Session 1 Historical Studies in Science Publishing
- Session 2 Contemporary Problems in Publishing the History of Science

No. 14 HISTORY OF SCIENCE: METHODOLOGY AND PHILOSOPHIES
- Session 1 Authors
- Session 2 Problems

No. 15 HISTORY OF SOCIAL AND HUMAN SCIENCES
- Session 1 Psychology
- Session 2 Social Sciences in the 19th Century
- Session 3 Social Sciences in the 20th Century
- Session 4 History of Geographic Thought

No. 16 SCIENCE AND RELIGION
- Session 1 Religion and the Scientific Revolution
- Session 2 Theology and Science in the 18th and 19th Centuries
- Session 3 Changing Views of Science and Religion

No. 17 CONTEXTS OF TECHNOLOGICAL CHANGE
- Session 1 Invention in Historical Perspective
- Session 2 Technology Transfer Across National Borders in the 20th Century
- Session 3 Technological Style
- Session 4 Technological Systems and Networks

No. 18 TRANSMISSION OF MATHEMATICAL SCIENCES
- Session 1 Networks and Dissemination of Mathematical Knowledge: Part A
- Session 2 Networks and Dissemination of Mathematical Knowledge: Part B
- Session 3 Transmission of Ancient and Medieval Mathematics
- Session 4 Transmission of Problems and Approaches: Part A
- Session 5 Transmission of Problems and Approaches: Part B
- Session 6 Mathematics in the Third World

No. 19 DOCUMENTATION
- Session 1 National Strategies for Preserving and Organizing Archives
- Session 2 Non-Written Sources in Historical Research

No. 20 HISTORICAL METROLOGY
- Session 1 Social Measurement
- Session 2 Physical Measurement

No. 21 SCIENCE AND TECHNOLOGY POLICIES AND DEVELOPMENT

Julio Samsó

Professor. Arabic Department. University of Barcelona
(Spain)

ALPHONSE X AND ARABIC ASTRONOMY

The relationship between Alphonsine and Arabic Astronomy will be studied in three different types of sources:

1. <u>Translations</u> : importance of translations of lost Arabic originals will be stressed. On the other hand, peculiar features found in these translations pose important problems of interpretation : such is the case of the abnormal situation of the equant (between the centre of the world and the centre of the deferent) which we can find both in Ibn al-Samh's <u>equatorium</u> and in the Latin Alphonsine version (<u>De cônfiguratione mundi</u>) of Ibn al-Haytham's <u>Hay'at al-ᶜalam</u>.

2. <u>Additions to translations and adaptations of Arabic works</u> : additions made by Alphonsine astronomers to their version of Qustā ibn Luqā's <u>Book on the Celestial Globe</u> imply an improvement of the <u>original instrument</u>. Additions to the <u>Libro de las Cruzes</u> can be due to the translator's awareness of the extremely crude astrological techniques used by the author of the original text. Al-Sūfī is the main source used by the compilers of the books on the <u>Ochava Esfera</u> but it is not the only one and there is a lot of material in them due to the work of Alphonse's collaborators. Work done recently on the two Alphonsine books on the planispheric astrolabe stresses the importance of the influence of the school of Maslama al-Majrītī.

3. <u>Original works</u> change, sometimes, our points of view on some questions. Such is the case of the book on the <u>Cuadrante sennero</u>, an observational instrument bearing a certain ressemblance to a double quadrant used in Marāgha. On the other hand this book bears witness to the competent use, by Alphonse's collaborators, of trigonometrical techniques developed in the Muslim East in the 10th-11th c. The <u>Alphonsine Tables</u> are the first <u>zīj</u> sponsored by a Christian king and a comparison in depth should be made between the Castillian <u>canones</u> and the Latin tables. Some questions of detail will be analysed in this last work such as, f. ex., the possibility of Sassanian influences in the solar model.

Jerzy Dobrzycki, Polish Academy of Sciences, Warsaw

The "Tabulae Resolutae"

From the very beginning the Alfonsine Tables in their Paris version have been repeatedly transformed to reduce the computational efforts of the astronomical practitioner. One way to simplify the structure of the tables was to provide explicitly the equations, thus avoiding multiple entries in the planetary tables. This was done already by John of Lignieres and repeated in various versions /cf. J.North, The Alfonsine Tables in England, Prismata, 1977 p. 269/. Another less sophisticated way was to leave the tables of equations unaltered and to simplify the chronology, computing ephemerides of mean motions, reduced to the local meridian. A very early example was a table of /Alfonsine/ Lunar mean places for 1300-1400 and the meridian of Paris, written alongside an earlier /Toledan/ table /MS Münich Clm 28229,13/.

In a developed form this type of tables became, as "Tabulae resolutae", a highly popular tool of the profession in Central Europe. A standard set of Tabulae resolutae consisted of two parts: tables of mean motions for the years 1428-1808 /anno completo/ and tables of equations, taken over unaltered from the Alfonsines. The specific selection of year dates suggests the influence of Oxford tables of 1348. Set up presumably in Prague, the T.resolutae came to Cracow early in the 15th century. Here they were introduced into the academic curriculum and, during the century, provided several times with successively more effusive explanations. Also the canonic set of tables was extended by the addition of various tables, mostly from John of Lignieres.

The Cracow T.resolutae were diffused and enjoyed some popularity in Europe in the late 15th century. They were even attributed to the Cracow authors /e.g. to Albertus of Brudzewo, Ox.Bodl. MS can.misc. 499/.

The concept and the name of the T.resolutae survived in later printed tables of Virdung and Schoner.

Juan Casanovas

Astronomer, Vatican City Observatory

On the Precession Problem in the Alphonsine Tables

   Although no numerical tables survive of the early version of the Alphonsine tables, from the "canones" in Spanish published by Rico y Sinobas, it can be deduced that the motion of the equinox and change of obliquity adopted were similar to the scheme proposed by al-Zarqallu.
   Jean de Murs in his commentary knows only the later Latin version where a compound precession is used and the change of the obliquity is ignored. He gives a few hints on how that apparently radical changes might have taken place.

Emmanuel POULLE

Ecole des chartes, Paris (Fr.)

Les canons des tables alphonsines au XIVe siècle.

On sait que, lorsqu'il est question, au Moyen Age, des tables alphonsines, il s'agit toujours de l'ensemble de tables astronomiques diffusées sous ce nom par les astronomes parisiens à partir de 1320, et non des tables établies sous le règne d'Alphonse X de Castille. Les conditions de l'élaboration de cet ensemble restent mal connues ; même la liste de ses composantes est difficile à établir car les tables astronomiques représentent un cas de tradition textuelle spécialement fragile et fluctuant. Mais les canons des tables, dont la tradition manuscrite obéit de façon plus classique à celle des textes littéraires, permettent d'appréhender, dans une certaine mesure, le contenu des tables dont ils décrivent le mode d'emploi.

On possède plusieurs canons des tables astronomiques rédigés dès l'introduction de celles-ci et dans les décennies suivantes, notamment par les astronomes parisiens qui sont responsables de leur diffusion : Jean de Lignères, Jean de Saxe, Jean de Murs. Chacun a produit son propre jeu de canons ; leur étude comparée peut apporter quelques informations sur la mise des tables alphonsines dans la forme sous laquelle elles ont ensuite été associées aux canons de Jean de Saxe.

On voit ainsi se préciser les caractéristiques techniques qui contribuent à définir ce que sont proprement les tables alphonsines: une sexagésimalisation systématique, avec son corollaire, l'affranchissement des contingences chronologiques, une application originale de la théorie du mouvement d'accès et de recès, un renouvellement partiel des équations, donc des excentricités ptolémennes, un programme tabulaire limité à la seule détermination des longitudes des planètes, à l'exclusion des latitudes et des conjonctions des deux luminaires.

Owen Gingerich

Harvard-Smithsonian Center for Astrophysics, Cambridge, MA, USA

THE ALFONSINE TABLES IN THE AGE OF PRINTING

The fame of Alfonso X in the Renaissance rested primarily on his astronomical tables, which were among the first scientific works available in the new age of printing. The Alfonsine Tables with the canons of John of Saxony were first printed in Venice in 1483, and, with the substitution of Santritter's canons, were reprinted there in 1492. It was this second edition that Copernicus owned and used extensively, particularly for its lunar tables and for its calendrical material.

The tables were printed again in Venice in 1518 and 1524, and in Paris in 1545. In addition, a variant of the tables with radices added for Queen Isabella of Spain was published under the title Tabule Astronomice Elizabeth Regine in 1503, and this form was included in some of the later editions. A somewhat more distant variant, the tabulae resolutae, also found its way into print in the tables of Johannes Schöner (1536) and Johannes Virdung (1541).

Probably the most important influence of the Alfonsine Tables in the century from 1470 to 1570 was not so much in the printed tables themselves as in the printed ephemerides generated from them, beginning with those of Regiomontanus in 1474. These were continued by Stoeffler and Pflaumen, and later by Pitati and Gauricus. Although most ephemeris makers promptly switched to the Copernican Prutenicae Tabulae after 1551, Alfonsine-based ephemerides were occasionally published, such as those computed by Cyprian Leovitius for 1556-1606. Tycho Brahe's dramatic adverse comparison of the Alfonsine predictions versus the Prutenics for the great conjunction of 1563 (a month's error vs a few days) turns out to be nontypical, as the Copernican improvement was generally far less apparent.

Indeed, the Alfonsines were still reprinted as late as 1647 (Florence), but by then they could hardly be considered a viable alternative to the new methods of planetary calculation.

Georg Bossong

Prof. Dr., Institute of Romance Philology, University of Munich, Ludwigstr. 25, D-8000 München (W-GERMANY)

SCIENCE IN THE VERNACULAR LANGUAGES: THE CASE OF ALFONSO X

In the history of science in Western Europe, the astronomical works of the Alphonsine School mark a turning point with respect to contents as well as to form. For the first time a Romance vernacular emerges as a medium of scientific thought. This had an impact not only on the development of Castilian and the other Ibero-Romance vernaculars but also on the evolution of European science in general. The rapid progress in theoretical and applied astronomy would hardly have been possible without the "linguistic democratization" of science performed by the learned king and his collaborators. The adaptation of the vernacular for scientific use created serious problems which were not only terminological. The resources of syntax had to be considerably enlarged, too. Algebraic formulae of the kind we are accostumed to began to be used only centuries later. The complex interplay of mathematical operations which makes up an equation had to be formulated in ordinary language. The solutions found to these problems reveal a profound influence of the Arabic models. To the translators and astronomers at Alfonso's court in Toledo, the scientific heritage of Greek and Old Indian Antiquity was transmitted in a linguistically arabicized form. It was this form which left its distinctive mark on the astronomical literature in Ibero-Romance until the end of the Middle Ages. - A general conclusion can be drawn. Linguistic problems play an important, although neglected, role in the history of science. Scientific thought needs appropriate linguistic means. Contents cannot develop independently of form. In the history of science, this is a chapter still to be written.

J.D.North
Professor of History of Natural Philos., Rijksuniv. Groningen

The Alfonsine Books and the Techniques of Astrology

Practising astronomers in the middle ages found especial difficulty with the whole range of trigonometrical problems of conversion as between equatorial and ecliptic coordinate systems. This was no doubt one of the reasons for their conservatism in the division of the ecliptic into 'houses' (the so-called 'mundane houses'). There are nevertheless as least seven different ways of effecting this division that are historically significant, and some of these are of great technical sophistication. The history of these methods has never been well understood. Those that were in the West assigned to Campanus of Novara and Regiomontanus, for example, are both known from earlier times. They, and others, are to be found in particular alluded to (in some cases very obliquely) in the Alfonsine books. Five of the books will be considered for evidence as to the use of the various methods, and for the names of astronomers associated with them--such as Ptolemy and Hermes (of course), al-Battani, Ibn az-Zarqellu, and al-Jayyani. The so-called Jaén tables offer a tentative link between Regiomontanus and the last-named scholar.

The trigonometrical problem of equating the houses is one which may be greatly eased with the help of an astrolabe, whether ordinary or universal. I shall introduce further evidence, in the form of an early fourteenth-century astrolabe from Granada, for the 'pre-history' of methods that came into their own only with the Renaissance.

This entire subject is one with strong implications for the analysis of horoscopes for the general historical evidence implicit in them. It is a subject treated in some detail in a recent work of mine, Horoscopes and History, in which the familiar methods of dating horoscopes from planetary positions are supplemented by methods indicating not only their styles but the probable geographical latitudes of their subjects.

Juan Vernet
Professor. University of Barcelona.

ALPHONSE X ET LA TECHNOLOGIE ARABE.

Bref exposé à propos des techniques arabes connues en Espagne et qui sont arrivées jusqu'à nos jours, suivi de l'étude de certains aspects du développement de la technologie hispano-arabe entre le Xe et le XIIIe siècles telle qu'elle apparait dans un manuscrit copié à la cour d'Alphonse X le Sage. Malgré le très mauvais état de conservation du manuscrit, on a pu reconstruire quelques machines qui y sont décrites. L'expérience acquise par l'équipe formé par M.V. Villuendas, R. Casals et J. Vernet permet de continuer ce travail avec l'espoir d'arriver à reconstruire, dans un futur prochain, la plupart des mécanismes.

Marjorie Nice Boyer

Professor emerita of History York College of the City University of
New York

The Late Antique and the Late Medieval Luxury Vehicle

In Western Europe luxury vehicles conferred prestige on their owners both in Late Antiquity and in the Late Medieval period. The two eras are separated by almost a thousand years, and during many centuries riding in a vehicle was forbidden to any self-respecting knight. There were the hangman's cart, the villein's cart, and the merchant's wagon, and, after 1200, the great lady's carriage. In none of these was there a place for a man of good family. Only about the middle of the fifteenth century did a luxury vehicle once again contribute to the dignity of a king or nobleman.

In Rome numerous mentions of vehicles first occur in the first century B. C. Under the Empire their use seems to have become more common and more prestigious. Perhaps the most glamorous were the racing chariot and the processional car in which emperors celebrated their triumphs, but elegant personal vehicles were much admired. The statues of the gods were paraded seated in chariots, and prominent men were given permission to decorate theirs with silver. On sarcophagi bas-reliefs of the deceased emphasized his social status.

The Merovingian kings were still riding in four-wheeled vehicles, but over the centuries their use for prestige or even for travel disappeared. During much of the Middle Ages the illustrator of a Biblical or astrological text had the choice of placing Elijah or Pharaoh or Sol in a chariot copied out of an ancient manuscript or in a contemporary farm wagon or dart. However, in the thirteenth century there was a change. During the last centuries of the Middle Ages a queen or countess required two carriages to maintain her state, one for herself and one for her ladies. The array was essential on ceremonial occasions. In the fourteenth century the suspended carriage was introduced into France. Perhaps because of the break in continuity between the ancient and the late medieval luxury vehicle there was little resemblance between the two, except for the rounded shape of the tilt. There were differences in the body and still more in the undercarriage.

Svante Lindqvist

Royal Institute of Technology, Stockholm, Sweden

A TWO TONS NUT & BOLT - THE RECONSTRUCTION OF A MEDIEVAL TREADWHEEL

Our opinion of medieval technology is shaped by those illustrations that have come down to us. They are often simple woodcuts or drawings, depicting the machines as rather rough-and-ready, unsophisticated constructions. The machine is reduced to a black-and-white, two-dimensional reproduction which fits between the covers of a book. It is hard to visualize its size, still harder the knowledge needed in order to build it. In the few cases where wooden constructions have survived, the wood is dark and cracked. Perhaps we unconsciously disparage medieval technology because the illustrations are so simple and the artifacts so poorly preserved. But what did the machine look like the day the carpenter put the finishing touches to his work, when the timber was fresh and smelling of resin ? Wouldn't our view of medieval technology change if we could once stand before such a construction ?

This was the background to the building of a full-scale replica of a treadwheel of 1520, which is preserved in the loft of Storkyrkan, the cathedral in Stockholm. The work was done in 1979 by students at the Royal Institute of Technology in Stockholm as a course project in history of technology.

The project gave a deeper knowledge of the wood-based technology of an earlier age, and in particular an insight into the know-how and skilled craftmanship of medieval engineers. In the course of our work, our perceptions of medieval technology changed. Life was suddenly breathed into the old woodcuts, and we saw them with new eyes.

Elspeth Whitney

Lecturer, Pennsylvania State University, Worthington Scranton Campus,
Dunmore, Pa.

SOME THIRTEENTH-CENTURY CONCEPTS OF THE MECHANICAL ARTS: ALBERTUS MAGNUS, THOMAS AQUINAS, ROBERT KILWARDBY, ROGER BACON

Whereas classical philosophers had for the most part regarded crafts as non-rational, servile, and degrading, twelfth- and thirteenth-century scientists and theologians began to give technology moral and intellectual sanction, either by placing the mechanical arts within the context of man's efforts to restore himself to his prelapsarian condition, as did Hugh of St. Victor and his followers, or, as in the Arabic tradition, by defining the intrinsic value of the mechanical arts in terms of practical science. In the thirteenth century, although the Victorine understanding of crafts as an aspect of salvation continued to be influential, the concept of the mechanical arts as applied science serving the community predominated. This emerging secular view of crafts, however, took diverse forms. Some thinkers, in particular, Thomas Aquinas, remained relatively untouched by contemporary intellectual currents on the status of the mechanical arts and closely followed Aristotle in emphasizing the inferiority of crafts to speculative science as a necessary consequence of their utilitarian nature. Other thinkers, however, attempted in different ways to fashion a coherent and positive view of technology from the diverse body of thought developed by their contemporaries and predecessors. Albertus Magnus, Robert Kilwardby and Roger Bacon, especially, modified elements taken from the Aristotelian, Arabic and Victorine traditions to define the mechanical arts as the operative or instrumental side of the theoretical sciences. These figures, whose thought on the mechanical arts has heretofore been treated largely in isolation both from each other and from the broader classical and medieval philosophical tradition on crafts, can usefully be seen as examples of a more general intellectual interest in articulating the status of technology as a category of knowledge. If they did not entirely overcome the pejorative attitude inherited from antiquity, they nevertheless made an important contribution to the Western cultural assimilation of technology.

Robert Fox

British Academy Reader in the Humanities, University of Lancaster, UK.

SCIENCE AND INDUSTRIAL RENEWAL IN FRANCE, 1851-1918

In recent years, assessments of the state of the French economy in the late nineteenth and early twentieth centuries have been swept by a strong revisionist current. As a result, it no longer seems adequate to speak of industrial stagnation across the board. For although there were sectors which palpably flagged, French industry had successes which make it essential to draw clear distinctions between sectors, periods, and regions.

These distinctions and the generally more sympathetic tone of the current secondary literature provide the starting point for my paper. My aim is not to present an apology for French industry, still less to conceal the special difficulties, political and cultural as well as economic, under which it laboured. But I do try to move behind the alarmist rhetoric of Albin Haller and certain other men of science who developed a dismal view of France's performance, especially in the realm of modern science-based industry, from about 1890. In answering a number of obviously central questions, I hope to move towards an appreciation of the finer structure of the problem. Why, for example, did France lose her initially strong position in the realm of dyestuffs (c.1860) and in electrical technology (c.1880)? Was education to blame, as many contemporaries maintained? Did the fault lie in the French character? Or was France just peculiarly vulnerable to the economic depression that affected the whole industrial world for much of the last quarter of the nineteenth century? And how do these and other explanations stand up in the light of the French successes (for instance, in the field of superphosphate fertilizers) to which our attention is now increasingly being drawn?

The answers offered in this paper are necessarily exploratory, arising as they do from the early stages of a wider study of the relations between scientific education and research and industrial performance in Europe now in progress at the University of Lancaster.

Samarendra Nath Sen

Ramakrishna Mission Institute of Culture, Calcutta, India

TECHNOLOGICAL EDUCATION IN INDIA : 1884-1914

The period 1884-1914 is of special significance in the history of technological education in India. It witnessed the establishment of several technical institutions, but, more importantly, intense debates on the scope, character, and relevance of such education. The Educational Despatch of 1854 referred briefly to professional education but was silent about technical education as such. The Education Commission of 1882 was struck by a complete neglect of useful and practical studies and, for the first time, introduced a technical bias by their concept of bifurcation of studies.

In India, technical and technological education were regarded as two parts of the same spectrum, the former being limited to the lower and middle level training and the latter to the advanced. In the beginning technical education embraced all forms of professional and specialized education, but was gradually applied more and more to education of scientific principles and processes affecting industries.

During the period in question three trends were visible - (a) emphasis on technical education at school and elementary level, (b) introduction of drawing and elementary science at primary and secondary stage as preparing the base of technical education, and (c) improvement of the advanced level through diversification of courses at engineering colleges and establishment of technological institutes. Most provincial governments laboured to produce elaborate plans more or less on European models, but practical achievement was disappointing in view of the absence of, or a deliberate policy against, a policy of industrialization along modern scientific lines.

Mudholkar refuted the patent government plea that technological education was not warranted by the state of existing industries and identified a large number of areas, - textile, mining, metallurgy, chemical industries, locomotion, mechanical and electrical engineering, where technological education held out immense possibilities. The point is illustrated by the case of textile technology. The concept of advanced technological education side by side with scientific education is discussed in the context of two newly founded institutions e.g., the Indian Institute of Science, Bangalore and the University College of Science and Technology, Calcutta.

Wolfhard Weber

Ruhr-Universität Bochum, Germany (FR)

GERMAN "TECHNOLOGIE" VERSUS FRENCH "POLYTECHNIQUE"
IN GERMANY 1780-1830

Das Bedürfnis nach einem verfügbaren Wissen über Technik ist von Francis Bacon theoretisch und von Jean-Baptiste Colbert praktisch umgesetzt worden. Doch die Konzepte zur gesellschaftlichen Einbindung eines technischen Wissenskanons entwickelten sich in Deutschland und Frankreich unterschiedlich. In Deutschland sollte es eher einer zentralstaatlichen Leitung die Übersicht erleichtern, in Frankreich unter dem Einfluß der Enzyklopädisten sollte es mehr den Vorstellungen eines politisch emanzipierten Bürgertums dienen. Beide Richtungen trafen nun in Berlin um 1795/1810 aufeinander und sahen sich zudem gemeinsam einer völlig neuen Idee von Universität gegenübergestellt. Die Erwartung an diese beiden Wissenschaften und ihr Durchsetzungsvermögen mußten daher sehr unterschiedlich sein.

Die "Technologie" war 1755/1772 erstmals in Deutschland als eine Verfahrenskunde zur Materialverarbeitung vorgestellt worden mit dem Ziel, den Beamten des Landesherrn Übersicht und Kontrolle zu ermöglichen. Demgegenüber erhob Gaspard Monge in seiner deskriptiven Geometrie den Anspruch, dem auf Mitteilung seiner konstruktiven Ideen bedachten Ingenieur eine unverwechselbare und eindeutige Sprache zu vermitteln, die zudem dem rationalen Ideal der mathematischen Berechenbarkeit unterlag. Zwar hat auch die Technologie in ihrer Form als Verfahrenskunde und mit ihrer starken Nähe zur Handwerksbeschreibung in Frankreich Anhänger gehabt (s. die Handwerksbeschreibungen Réaumurs), die in den Jahren der jakobinischen Herrschaft und Forderung nach unmittelbarer Anleitung zu gewerblicher Tätigkeit durch Hassenfratz befürwortet wurde, doch hat sich schließlich für die Erfordernisse der Konstruktion die auf der Ecole Polytechnique fortentwickelte Geometrie als Voraussetzung aller weiterführenden Ausbildung erwiesen.

Die Idee der Polytechnique zog mit Napoléon durch Europa, aber sie siegte nicht in Berlin. Verstärkt nach 1830 empfand man in Preußen das Polytechnische in seiner Rechenhaftigkeit als etwas "Amerikanisches", das man verabscheute und in die Niederungen der gewerblichen Ausbildung abschob. Dort, in der Kompetenz des Handelsministers, war man unter Aufnahme polytechnischer Anregungen aber auf dem erfolgreichen Weg zum Aufbau einer eigenen eher praxisorientierten Maschinenbauschule und näherte sich schnell dem anspruchsvollen Niveau einer höheren technischen Bildungsanstalt, auch ohne den politisch so gefährlichen Begriff "polytechnisch" zu verwenden.

Stefan G. Balan

Member of the Romanian Academy

SCIENCE, TECHNOLOGY, EDUCATION

1. Science, technology, education. The science about the existence consists of a systematic assembly of veridical theoretical knowledge about nature, society and thinking. Technology is a form of active manifestation of man on nature, through means of production, methods of work and specific knowledge, with a view to create or improve material goods or services. When, in the course of time, man understood the interdependence between technique and nature sciences the synthesis of nature sciences and technology begun, to the extent this was necessary and possible, for the progress of mankind. Thus the technological sciences were established.

We must note that in whatever is newly achieved, theoretical knowledge, older or more recent, from the sciences of nature, is "absorbed", the same as some knowledge, theoretical or practical, from the technical or general sciences. This knowledge helps the inventor to achieve what he thinks, the volume of scientific knowledge increasing man's force of action on nature. Understanding the importance of possessing ample knowledge from other sciences, through learning, in order to increase the technological level in the life of human society, more and more attention has been given to education. Education has become a chain and link between the knowledge from the nature sciences and the knowledge from the technical sciences. The technician, in order to make sensible and rapid progress, must be well experienced, through practical and theoretical learning, in the nature sciences, in the general sciences (mathematics, informatics etc.) and of course in the technological sciences in which he works.

2. In the second part of the paper examples are given of the link between science and technology through education, from the main periods of development of Romania. In antiquity there were few schools, at the apprenticeship level; the tehnology was developed in some fields, such as mining, primary metallurgy, wood, building, agriculture; later the schools appeared (the oldest one in Cenad, in Latin, in the year 1020); in the Middle Ages the technology develops mostly in the fields of mining and building. The printing press appears. When the universities and the higher technical schools were set up, the technology also developed. At the same time the economy developed.

Thus the importance of education in the triad science - technology - education is demonstrated practically.

LIVIU SOFONEA

University of Brasov   Rumania

Some historical and epistemological aspects of technology training and education in Romania

In the paper some facts, lines, trends, and comments are proposed in the history of Romanian schools, new dictionaries concerning the history of science and technology, the popular technologies (museums, exhibitions, reconstructions, present use, pedagogical aspects, etc.), and historical and epistemological models concerning some direct and indirect connections between science-technics-technology and society (the case of energy; ecological-axiological aspects, etc.).

A part of the paper is devoted to older as well as recent researches concerning the explanation and interpretation of the ancient computing system in Dacia. In the mountains of Transylvania, in the preroman captial of Dacia named Sarmizegetusa Regia, some archaelogical monuments have been discovered and plausibly interpreted as sanctuaries (big and small, round and rectangular ones) where the calendar was made and recorded. An exegesis of all previous researches (numerological as well as astronomical) is presented, together with some original contributions. Other trends are also emphasized.

Alexandre HERLEA

Conservatoire National des Arts et Métiers, Paris, France.

ENSEIGNEMENT TECHNIQUE SUPERIEUR ET LABORATOIRES DE RECHERCHE INDUSTRIELLE EN FRANCE AU XIXe SIECLE : L'EXEMPLE DU CNAM.

L'enseignement supérieur en général et l'enseignement technique en particulier furent partout et depuis leur naissance intimement liés à la recherche fondamentale et appliquée. En France le Conservatoire National des Arts et Métiers (CNAM) est au XIXe siècle un des exemples les plus caractéristiques, sinon le plus caractéristique des relations étroites existant entre le développement de l'enseignement technique supérieur d'une part, et celui de la recherche orientée vers l'application industrielle d'autre part.

Ce caractère, particulier au CNAM, ne cessera de s'amplifier avec l'établissement d'écoles spécialisées (dessin 1799, filature 1804) et des chaires d'enseignement : 3 en 1819 (mécanique, chimie, économie), 7 de 1829 à 1839, 3 de 1848 à 1854, 7 de 1879 à 1900. Ces chaires correspondent à la création des nouvelles disciplines apparues au fur et à mesure du progrès technique. Les chaires sont établies le plus souvent "ad personam", se dotant ainsi de la compétence d'un spécialiste de prestige, voire d'un créateur de discipline nouvelle. Ces spécialistes poursuivent leurs recherches dans le cadre de laboratoires spécialisés crées auprès de leur chaire, et qui jouent plusieurs rôles : laboratoire de recherche ; laboratoire pédagogique pour les travaux pratiques des étudiants ; laboratoire d'essai pour l'industrie.

Parmis les chaires et les laboratoires les plus célèbres du CNAM, mentionnons ceux de mécanique, métrologie, physique, chimie, agriculture, électricité, métallurgie, etc...

Notre communication porte sur l'ensemble de ces laboratoires dans leurs relations avec les chaires. Résumons ici à titre d'exemple ce qui concerne la physique et l'électricité.

Claude S.M. Pouillet, directeur du CNAM et titulaire de la chaire de physique créée pour lui en 1829, évalua pour la première fois la chaleur solaire à l'aide de son pyrhéliomètre, inventa le pyromètre magnétique pour basses températures et établit, indépendamment d'Ohm, les lois fondamentales des courants électriques grâce à des instruments de son invention. Plus tard, dans son laboratoire alors dirigé par Alexandre E. Becquerel, Gaston Planté inventa l'accumulateur électrique. C'est toujours au CNAM que fut créée en 1890 la première chaire d'électrotechnique pour Marcel Deprez, à qui on doit la théorie et les expériences décisives qui ont permis la transmission de l'énergie électrique à longue distance au moyen des lignes à haute tension ainsi que plusieurs appareils de mesures (galvanomètre astatique etc.). Dans son laboratoire Henri Moissan créa le four électrique qu'il utilisa pour la préparation des aciers spéciaux, la synthèse du carbure de calcium etc...

Wolfgang König

Verein Deutscher Ingenieure, Düsseldorf, Bundesrepublik Deutschland

Science and Practice: Key Categories for the Professionalization of German Engineers

In international comparison we can work out three distinctive aspects of the German system of engineering education: 1) a strict separation between a higher ("Technische Hochschule") and a lower branch ("Technische Mittelschule"/"Fachhochschule"), which resulted in different professional careers of the graduates: 2) a strong practical orientation of the education, differently developed in the two branches; and 3) strictly formulated curricula in both branches.

These specific aspects are results of a historical development over more than 150 years. The older literature in particular has dealt with the scientific development of the "Polytechnische Schulen" and their maturing into "Technische Hochschulen" in the 19th century and the way they were increasingly fashioned after the model of the universities (Karl-Heinz Manegold). Increasingly scientific orientation ran in part counter to the needs of industry (Jürgen Kocka). However, there are other tendencies that have not been investigated to the same extent: 1) the development of a second branch of engineering education by the foundation of the "Technische Mittelschulen", and 2) the successive integration of elements of practice in the "Technische Hochschulen."

It was industry and the Verein Deutscher Ingenieure that had an important influence on this development, which was the decisive moment for the formation of the occupational group of the engineers. This formation did not come to an end until the passage of the engineering laws at the beginning of the 1970s.

Dr. Hans-Joachim Braun

Professor, University of the Federal Armed Forces, Hamburg, School of Education

TECHNOLOGICAL EDUCATION AND TECHNOLOGICAL STYLE IN GERMAN MECHANICAL ENGINEERING; 1850-1914

For a latecomer in industrialization technological education seemed to be an appropriate means to foster industrialization and economic growth. In mechanical engineering neither the model of cameralist technology with its bias towards empiricism nor technological education à la Français (Ecole Polytechnique) seemed to be the right way to achieve that end. At Karlsruhe Polytechnic, founded in 1825, Ferdinand Redtenbacher established the subject Maschinenwissenschaft, incorporating traditional German and recent French elements and stressed that this new subject was more than applied science. Redtenbacher's model was widely followed in Germany. In the 1870s and 1880s mechanical engineering Professors like Franz Reuleaux and Franz Grashof brought mechanical engineering closer towards mathematics and the natural sciences with the intent of raising the social prestige of this subject and its practitioners. This kind of theoretical mechanical engineering education raised, however, widespread complaints in German industrial circles. Many industrialists and mechanical engineering professors like Aloys Riedler demanded a more practical bent with the foundation of engineering laboratories and practical education of engineering students on the shop floor. At the turn of the century various German Institutes of Technology responded positively to these demands. After this, it is possible to speak of a particular German technological style in both higher technological education and research and development work in mechanical engineering corporations.

Robert Angus Buchanan

Centre for History of Technology, Science and Society, University of Bath, U.K.

EDUCATION OR TRAINING? THE DILEMMA OF BRITISH ENGINEERING IN THE NINETEENTH CENTURY

The consensus of scholarly opinion has hitherto been virtually unanimous in condemning the British engineering profession for the manifest reluctance with which it adopted academic education as part of its admission procedure in the nineteenth century. Exponents of technical and scientific education, in particular, have frequently been vehement in their criticism of what they have seen as professional conservatism. But there is another side to this story which deserves to be recorded, because British engineering was faced with a genuine dilemma in the second half of the nineteenth century. Riding the crest of rapid industrialization and the boom in railway construction which had created an unprecedented demand for their services, the engineers were a successful and rapidly growing profession. With hardly any exceptions these engineers had acquired their skill by a system of "in service" training or pupilage: they were attached to the office of a practising engineer for three or four years on payment of an annual premium, during which time they learnt as much as they could of the theory and practice of engineering. Thereafter, they kept their knowledge up to date through the medium of their professional institutions which fulfilled an important educational function.

This system of training by apprenticeship had served British engineering outstandingly well, and was not to be lightly abandoned. On the other hand - and here the dilemma becomes apparent - it could not provide an adequate basis of theoretical knowledge in the new technologies: electrical engineering, chemical engineering, thermodynamics, etc. which became increasingly important in the second half of the nineteenth century. The solution to this problem was a diplomatic compromise. British engineering retained its strong attachment to practical experience, but it combined this with a recognition of the vital role of an academic education in providing the theoretical knowledge which had become essential to all branches of engineering. The professional institutions, moreover, have preserved and even extended their distinctive educational service, thus ensuring that their members are kept fully informed of new developments in their field of engineering.

Melvin Kranzberg, Georgia Institute of Technology

BROADENING AND DEEPENING U.S. ENGINEERING CURRICULA

Engineering education in the United States is broadly surveyed in terms of changing technical needs and emphases, the development of engineering professionalism, and the role of non-technical subjects in the engineering curriculum.

New engineering problems and methodologies were reflected in both professional institutions and education throughout the 19th century. Formal schooling of engineers superseded the apprenticeship type of training, when the Federal Government established the Land Grant agricultural and mechanical colleges during the 1860s and later. At the same time new fields of engineering were emerging from the older civil engineering, requiring different kinds of technical expertise, and the basic educational approach also changed from the "shop culture" to the "school culture."

These processes have accelerated during the 20th century; and new social attitudes and problems have emerged. Although at first these latter had but a short-lived impact on the engineering profession, following World War II wholesale changes in engineering were introduced as a result of new technical and social needs.

When, some two decades ago, the benefits of technology to society began to be questioned, engineers were bewildered. This response of the engineering professional community is examined, along with the emergence of new elements in engineering education. These include the role of social-humanistic studies; the consideration of science-technology relationships; the development of paraprofessionals (engineering technology); the viability of current institutional mechanisms in engineering education, including research partnerships with industry; the rise of ancillary studies in Science-Technology-Society (STS) programs; and the enlargement of traditional liberal arts education to encompass "Technological Literacy."

Jean Michel

Ecole Nationale des Ponts et Chaussées

Les Grandes Ecoles d'ingénieurs et la maîtrise du développement scientifique, technique et industriel en France

Les caractéristiques des Grandes Ecoles françaises consistent en une démarche pédagogique ("macropédagogique" en fait) originale. Cette dernière s'inscrit dans une perspective, très moderne, de maîtrise collective des savoirs scientifiques et techniques des groupes d'ingénieurs français. Peu d'analyses ont été faites pour comprendre les relations (l'interaction) entre ce système éducatif original et les développements scientifiques, techniques, industriels du pays.

La communication insistera sur ces aspects mal connus, en essayant de présenter successivement:

- la généalogie, la typologie et les traits caractéristiques du système des Grandes Ecoles: les réponses quantitatives et qualitatives aux besoins en ingénieurs; les Corps d'ingénieurs, la professionalisation de la formation, la dialectique "centralisme-antonomie", les liens avec l'industrie, etc...
- l'originalité du project (macro) pédagogique des Grandes Ecoles d'ingénieurs et l'organisation d'un système cohérent et global de création, de développement et d'entretien des savoirs spécialisés et spécifiques des ingénieurs: l'adaptibilité aux besoins, la démarche de finalisation, le rôle des systèmes d'information-formation, la recherche et la formation continue, etc...
- les conséquences de cette organisation du "système" des Grandes Ecoles sur le développement industriel français: la maîtrise de l'innovation technique, la conduite des grands projets (chemin de fer, espace, énergie nucléaire, informatique et télématique, métropoles urbaines...), l'adaptabilité des ingénieurs face aux évolutions de la société.

Arthur Donovan

Virginia Polytechnic Institute and State University

ENGINEERING EDUCATION AND THE PROFESSIONALIZATION OF ENGINEERING IN THE UNITED STATES

To what extent and in what ways is the area of occupational specialization called engineering a profession? During the past 150 years prominent non-military engineers in the United States, as in Great Britain, have sought to attain for engineering the high social status accorded to the older paradigmatic professions of the religious ministry, the law and medicine. In doing so they have emphasized the increasing social importance of technical knowledge, the undeniable foundation of engineering expertise, and they have created both a comprehensive set of professional societies and a highly structured system of specialized education. Many students of the rise of engineering have interpreted these developments as fairly straightforward steps in the professionalization of engineering.

During the past two decades, however, social historians have developed a different approach to the study of professionalization. These historians emphasize the ways in which the ideology of professionalism has been used in struggles over social status and social authority, struggles that are endemic in societies being transformed by economic and social change. In their accounts technical expertise is presented not as the substance of professionalism, but rather as one of the means by which the groups that possess such knowledge attempt to secure for themselves a favorable social position. Such a view renders the earlier accounts of the professionalization of engineering problematic.

In this paper I suggest that the history of engineering education in America needs to be re-examined, with special attention being given to the tensions between liberal education and professional specialization in American higher education. A social history of engineering education is called for, one that describes the development of engineering education with reference to both the social and career experiences of engineers and the changing social values of the universities within which engineering education came to be institutionalized. Such a study will help clarify the complex historical relationship between engineering and professionalism in America.

Dr Ditta Bartels   Dr W.R. Albury

Honorary Visiting Fellow Associate Professor

School of History and Philosophy of Science
The University of New South Wales, Australia

Recombinant DNA - The Politics of Scientific Innovation

   The control of scientific innovation in molecular biology has been widely recognised as a political issue since the early 1970s, when recombinant DNA techniques started to become available and researchers' concerns were first publicly expressed about potential hazards from genetically engineered organisms. In the U.S.A. and a number of other Western countries, such as Australia, public participation in the control of recombinant DNA research has been advocated from a variety of perspectives. Within the research community, however, attempts by non-experts to involve themselves in the regulation of scientific practice have generally been resisted on the ground that the advancement of knowledge requires freedom from political interference.

   Our discussion will examine some of the principal justifications which have been advanced for regular public participation in recombinant DNA regulation, ranging from idealised theories of democratic pluralism to pragmatic considerations of legal liability and commercial development. Our aim will be to assess the political case for this form of public involvement in science from the point of view of its advantages or disadvantages for the production of scientific knowledge.

Jonathan Harwood

University of Manchester, UK.

BREADTH VS. SPECIALIZATION IN INTER-WAR GENETICS

Ben-David has demonstrated the importance of national differences in institutional structure for the rates of science's growth in those countries. He has also shown the effect of academic structures upon specialty-formation. This paper goes further to argue that the intellectual content of such specialties – in particular their theoretical scope – is also shaped by their institutional setting.

German and American geneticists defined their discipline rather differently during the 1920s and '30s. In the US Morgan and his school deferred the more complicated questions of heredity's role in development and evolution until the simpler issue of the structure and transmission of the hereditary substance had been solved. In Germany, by contrast, many geneticists insisted that an adequate theory of heredity must address <u>all</u> of these problems. The marked interest in cytoplasmic inheritance in Germany (and its neglect in the US) reflected these contrasting conceptions of genetics' scope.

But genetics was not the only new discipline whose American practitioners embraced specialization more avidly than their German counterparts. An adequate explanation must, therefore, be systemic. One explanation for such 'national styles' can be found in the different institutional settings in both countries in which new disciplines were likely to develop. Economic crises in Germany after World War One certainly stifled new academic initiatives. But even before the war contrasting structures of their university systems tended to foster specialization in the US while hindering it in Germany. While much genetic research was conducted in Kaiser-Wilhelm Institutes, these did not provide long-term career possibilities for more than a few geneticists. By failing to establish itself outside the universities (thus remaining institutionally dependent upon its parental disciplines, botany and zoology), German genetics retained into the 1940s a theoretical breadth which had been jettisoned a generation earlier in the US.

M. J. S. Hodge

Philosophy Department, University of Leeds.

GENERATION AND THE ORIGIN OF SPECIES FROM DARWIN (1837) TO DOBZHANSKY (1937): HISTORIOGRAPHIC PROPOSALS

Two questions still dominate historiography of "cytology, genetics and evolution". How did cytology contribute to the "hard", "particulate", "non-saltationist" heredity of the "new synthesis"? How was Evolution, without Genetics up to 1900, then changed by "the rise of this new science of genetics"? - questions usually answered by contrasting "Mendelians" and "biometricians", "naturalists and experimentalists", etc. etc. However, much recent scholarship (e.g. Bowler, Churchill, Farley, Mayr, Olby, Sloan) shows that this historiography is not fully adequate.

The generation theories in Darwin's thinking suggest a new start. In 1837, before natural selection or pangenesis, his evolution theorising began with growth and asexual generation vs. sexual generation. By 1868, pg. and ns. were integrated publicly into a single structure of argument that, like Spencer's and Haeckel's, moved all the way from tissue growth and reproduction to the tree of life. Let us ask, then, how changes in cytological theory contributed to changes in such generational-evolutionary theorising after the 1860's. Take two cases: (a) Weismann and (b) Muller and Wright on sexual reproduction and evolution. Both show that, even with "hard", "particulate", "gradualist" theorists, the bearing of cytology on evolution is far from fully captured by the traditional historiography.

Natasha X. Jacobs

Doctoral Candidate at Indiana University

"THE MENDELIAN CELL AND OTHER MODELS: THE CYTOLOGICAL BASIS OF WEIMAR GENETICS"

The predilection of German biologists for genetically odd organisms culminated in the divisive, controversial views and programs of Weimar geneticists. But behind the facade' of competing theories of heredity, geneticists and cytologists translated confusion and conflict into highly productive research. Men like Carl Correns, Max Hartmann, Karl Belar, Otto Renner, Fritz Baltzer, and Erwin Baur, sought to determine structural and mechanical differences between the cells of organisms exhibiting Mendelian hereditary patterns and those in which certain traits (sex, coloration) did not mendelize.
Confronted by nature's baffling provision of normal and anomalous cytogenetic systems, Weimar biologists ultimately forced genetics to move beyond rather than against T. H. Morgan's analysis and theory of Mendelian inheritance.

D-r Musrukova E.B.

Institute for History of Science and Technology,

Moscow, USSR

Cytogenetic and Cytology in Russia between two world war

The development of Soviet cytology is closely related to those trends in research which formed at that period in Petrograd /Leningrad/ and in Moscow. In Petrograd cytological research was traditionally carried out by the Medical Surgery Academy /Medical Military Academy/ at the Chair of Histology and by the Chair of Zoology of the Petrograd University.

Cytological and cytogenetic investigations in Moscow were performed at the Moscow University, at the Institute of Experimental Biology and at the Chair of Histology of Medical Institutes. For instance, A.G. Gurvich was the Head of the Chair of Histology at the 1st Medical Institute. He was a prominent histologist and the founder of the theory of "biological field" /1923/.

Cytogenetic investigations were initiated by S.G. Navashin, the founder of karyology who worked in Moscow since 1923. He founded a school of national cytogenetics.

N.K.Kol'tsov is justly considered to be the founder of the Moscow school of cytology. Roskin, Lavrentiev, Rumyantsev and Chruschev were his pupils and worked with him. Soviet cytology takes pride in their investigations.

JAN SAPP

University of Melbourne

Discovery or Fraud: Franz Moewus and the Foundations of Microbial Genetics.

Scientific fraud is an important problem for historians and sociologists of science as well as for scientists themselves. Whatever the reasons, members of almost every major academic institution have only recently "discovered" fraud and realized that they have to deal with it in some sensible way. Yet our knowledge of fraud to date consists of limited information which is varied and unsystematic. In most cases almost nothing is known about the circumstances in which frauds occur, or indeed, what constitutes a fraud in science. The prevalance of fraud and its deeper significance remains largely an area of speculation and controversy. For historians and sociologists, charges of fraud, and its actual occurrence pose major problems for our conception of the nature of scientific activity, the social conditions for the progress of reason, and the way in which knowledge claims are assessed and certified. In this sense, fraud is the mirror image of discovery and a study of the controversy surrounding the German biologist Franz Moewus provides a case study for our more general interest in the nature of scientific discovery.

Franz Moewus stood at the threshold of a major revolution in biology and he anticipated, participated in, and made generous novel contributions to the published literature of modern microbial and biochemical genetics. Although it has generally escaped the notice of historians of genetics, Moewus was one first biologists to domesticate micro-organisms for genetic use. As early as 1940, a year before the celebrated paper by George Beadle and Edward Tatum appeared, Moewus published a series of papers in which he showed how biochemical analysis of <u>Chlamydomonas</u> and other micro-organisms could be done. Throughout that decade he provided and articulated a plausible biochemical model of how genes effect the control of the complete life-cycle of the organism.

Moewus' work rose to the center of a great deal of controversy during the decade following World War II. Some geneticists claimed that he had laid the foundations of microbial and biochemical genetics, and that he should be given priority recognition over Beadle and Tatum who were ultimately awarded a Nobel prize in 1958. On the other hand, Moewus was relentlessly criticized and defamed by many others and he was ultimately generally dismissed as the perpetrator of one of the most ambitious frauds in the history of science.

Diane B. Paul

Associate Professor, University of Massachusetts at Boston

THE "REAL MENACE" OF THE FEEBLEMINDED: SCIENTISTS AND STERILIZATION, 1917-1930

In 1917, E.M. East asserted that a policy of segregating or sterilizing the feebleminded would have little effect on reducing their numbers since most of the genes responsible for the trait were hidden in apparently normal carriers. The "real menace" of the feebleminded, he argued, was constituted by this large, and currently invisible, heterozygotic reserve. East's figures were refined, and conclusion strengthened, by R.C. Punnett who concluded that at least 10% of the American population carried the defective recessive gene for feeblemindedness. Policies aimed at the affected themselves would, he thought, be even slower in their effect than East anticipated. Punnett's pessimistic conclusion was popularized by the American geneticist H.S. Jennings. But it was challenged by R.A. Fisher, who noted that it was based on the unrealistic assumptions of random mating and single gene inheritance.

This controversy has sometimes been taken to exemplify the differences between "mainline" and "reform" eugenicists. However, the issues debated by the geneticists did not include policy toward the feebleminded themselves (whom all the participants advocated segregating or sterilizing) or the course of further research (which all the participants wished to direct toward identification of carriers). I will argue that this particular controversy better illustrates the assumptions shared by geneticists than it does their differences. It serves also as a good example of the social plasticity of scientific doctrines.

William Schneider

Associate Professor, University of North Carolina, Wilmington

Puericulture and the Style of Eugenics in France

Eugenics was a widespread concept throughout Europe and the United States in the beginning of the twentieth century. As shown elsewhere (Schneider, "Towards the Improvement of the Human Race," Journal of Modern History , 54, 1982) France was among those countries where it gained currency among a variety of groups, but when eugenics became an organized movement there, its leading French theorists were the most innocent and least threatening people imaginable: baby doctors. Even the name they used for the concept was changed from the harsh, Greek-derived "eugenics" to the softer word of Latin origin, "puericulture." Broadly defined as, "knowledge relative to the reproduction, the conservation and the amelioration of the human species," puericulture remained a key element of French eugenics long after its proponents relented and adopted the more common international term for the name of their organization: the French Eugenics Society.

Puericulture had spread widely and rapidly after its introduction by Adolphe Pinard in 1895, even gaining an international following, but it would have been little more than a call for better prenatal care and breastfeeding without its hereditary underpinnings. This so-called "puericulture before procreation" connected the well-being of the infant not just to the health of the mother but also to previous generations and those yet unborn. The concept was, therefore, surprisingly close to the eugenics of the Anglo-Saxon countries, but its style of presentation reflected the cultural setting of turn-of-the-century France.

One substantive feature of hereditary theory in France which reinforced this style of French eugenics was the predominance of neo-Lamarckism. The presumption that environmental influences could be inherited by subsequent generations made French eugenics in principle a very open and inclusive movement which could work with all whose goal was the improvement of the well-being of the population. Thus, with infant health as the sympathetic focus drawing many diverse interests to puericulture and French eugenics, neo-Lamarckian heredity provided the theoretical link that held them together.

Paul Weindling

University of Oxford, Wellcome Unit for the History of Medicine, U.K.

Racial Biology in Nazi Germany: a Divided Science

Nazi racism resulted in some of the worst ever scientific crimes against humanity. Yet the development of racial biology before 1933, and the research aims and organisation of biology during the Third Reich have received only limited historical attention. This paper looks at key eugenic research institutions like the Deutsche Forschungsanstalt für Psychiatrie and the Kaiser Wilhelm Institut für Anthropologie, as well as university institutes and the German Society for Racial Hygiene, in order to assess scientific developments, relations with the state and Party, the popularisation of racial biology, and finally links between biology and the sterilisation programme, euthenasia and genocide. I shall suggest that the situation was far more complex, and changing, than has hitherto been appreciated.

Nazi racial ideology confronted scientists with two major problems: that of racial purity and the inheritance of degenerate characters, like diseases or psychological traits. Prior to 1933 most racial anthropologists did not consider that there was a pure Germanic race, but saw a mixture of types e.g. alpine and nordic. Nazi concepts of Aryan racial purity and human heredity caused much dissension among anthropologists. The inheritance of diseases and of degenerate characteristics, as well as preventive measures to protect the health of future generations, were divisive issues among eugenicists. The situation was further complicated by groups in competition for power among the state and Party, so that it was possible for biologists not to have the favour of e.g. Himmler, and yet to find a protected niche under the patronage of others. The questions arise, whether eugenicists were able to maintain their position as arbiters of social policy as during the Weimar Republic, or whether eugenicists had only a subordinate role with the major initiatives being taken by the Party and state. De-nazification proceedings suggested that scientists were under considerable pressure to acquiesce in racial policies, so as to protect the integrity of their institutes. Reference will be made to Eugen Fischer, Nachtsheim, Rüdin, Verschuer, and Mengele. How racial biology was demarcated from human genetics sheds light on issues of science and ideology. Comparison will be made to British and US eugenics, and the question of whether there really was a new distinctive ethos of Nazi racial science will be posed.

Mark B. Adams

University of Pennsylvania, Philadelphia, USA

SOVIET MEDICAL GENETICS IN THE 1930S

In the 1930s, Soviet medical genetics emerged as a distinct field in Narkomzdrav's Institute of Medical Genetics. Although its research agenda and personnel were largely inherited from the Russian eugenics movement of the 1920s, its character and mission were shaped by three Marxist enthusiasts: A. Serebrovsky (1892–1948), American geneticist H. J. Muller (1890–1967), and S. G. Levit (1896–1943), a physician and active Party member who converted from Lamarckism 1927/28.

In late 1928, Serebrovsky and Levit organized a kabinet on human heredity and constitution within the Biomedical Institute to map human chromosomes and study human population genetics and pathology. Serebrovsky's lead article in its first publication (1929) advocated artificial insemination of women with the sperm of eugenically selected male donors. His plan was widely ridiculed during the Great Break (1929–32) for its elitism and "biologization" of social issues. In 1930, Levit became director of the Biomedical Institute and its Genetics Division, forcefully arguing the difference between eugenics and the new "anthropogenetics." Levit studied with Muller in Texas 1931/32; in January, the institute was suspended, but in September 1932 it reopened with a new mission (the genetics of human biology, pathology, and psychology), new divisions of cytology, experimental pathology, biometrics, psychology, and physiotherapeutics, and an expanded staff.

In 1933, Muller came to work in the USSR and became active in Levit's institute. Its conference on "medical genetics" (15 May 1934) was attended by 300 people and featured speeches by Levit, Muller, Kol'tsov, Davidenkov, Iudin, Bunak, and Andres urging the rapid expansion of the field to combat fascism and improve health. Its final resolution called for the creation of new clinics and medical school departments, more staff and graduate students, the preparation of texts and teaching materials, and new genetics courses for physicians. In March 1935, the institute was renamed the Maxim Gorky Scientific Research Institute of Medical Genetics. Its impressive fourth volume (1936, 540 pp.) reported work by thirty researchers, notably a sophisticated series of twin studies (including one by A. R. Luria, who had become head of the institute's psychology section).

In May 1936, Muller resurrected Serebrovsky's discredited breeding scheme in an letter to Stalin which proved untimely. Within months, the scheduled 7th International Genetics Congress (Moscow 1937) was postponed indefinitely because of its section on human genetics. The politics of human heredity helped Lysenkoists score their first major success at the VASKhNIL meeting (December 1936). Muller left Russia in early March 1937; within days Levit was arrested and the Institute was disbanded. Repressed at roughly the same time and for the same reasons, genetics, medical genetics, and eugenics were reborn together after 1963. Current Soviet controversies over nature and nurture have been largely shaped by this special institutional and ideological history.

Gunnar Broberg

Assoc. prof, Dpt of History of Science and Ideas, Uppsala univ.

EUGENICS IN SWEDEN DURING THE FIRST HALF OF THE TWENTIETH CENTURY

   Sweden followed the general pattern of interest in eugenic reform during the first decades of the twentieth century. A national society for eugenics (rashygien) with several prominent academic members was founded in 1909. Turning to science as a means to improve postwar society the Swedish Diet unanimously decided in 1921 to found a national institute for race biology; it was headed by Herman Lundborg up to 1935 when he was succeeded by Gunnar Dahlberg. The institute boasted to be the first federal institute of its kind in the world. Although it enjoyed considerable public interest, the institute was no success. Because of the developments in Nazi Germany scepticism grew; the undiplomatic Lundborg made matters worse. True, Dahlberg, radical, mathematically instead of biolocally oriented and adherent to environmental thinking, was very much his opposite. Thus, at a time of growing failure Swedish eugenics experienced a change in direction shortly before world war two.
   In many respects Sweden reflected the general development of eugenics. There were however some national features: the Swedish descriptionalist-Linnean tradition, an over-all ambition to map the national resources, organic as well as inorganic, and the hopes fostered by plant-breeding in Southern Sweden for the use of science in modernizing society.

Daniel J. Kevles

Professor of History, California Institute of Technology

HUMAN GENETICS: THE PENROSE SCHOOL

In 1945, Lionel S. Penrose was appointed the Galton Professor of Eugenics at University College London. Penrose was a leading authority on the genetics of mental deficiency, and he was moving into the more general area of human heredity. From his work in mental deficiency, Penrose realized that the study of human heredity required knowledge of statistics, biochemistry, and medicine as well as genetics, and, with the aid of J.B.S. Haldane, he institutionalized this multidisciplinary approach to the subject at the Galton Laboratory. Between 1945 and 1965, when Penrose left the professorship, the Galton became a mecca for human geneticists everywhere and did a great deal to foster the development of their discipline.

Peter J. Bowler

Lecturer, The Queen's University of Belfast, Belfast BT7 1NN.

HUMAN GENETICS AND EVOLUTION: R. RUGGLES GATES AND RACE THEORY

A significant minority of early 20th-century palaeoanthropologists supported a polygenist view of the human races. Invoking the principle of parallel evolution, Hermann Klaatsch, Earnest Hooton and others suggested that the races had separate origins and might be classed a separate species. They appealed to paleontology for evidence that parallel evolution did occur, but did not specify the evolutionary mechanisms that might be involved. At the same time, a few geneticists supported the unorthodox view that parallel mutations might give rise to non-adaptive orthogenesis. In 1948 R. Ruggles Gates tried to bring these movements together in his book <u>Human Ancestry from a Genetical Point of View</u>. He derided the sterility criterion and insisted that parallel mutations had produced several distinct human species that could interbreed. Gates' efforts came just as the Modern Synthesis was undermining the plausibility of the non-Darwinian mechanisms of evolution. The timing of his book suggests that the anthropologists had failed to communicate even with those geneticists who might have helped them to defend their non-Darwinian ideas.

Charles Galperin

Maître de conférence, Université de Lille, France

THE HISTORY OF LYSOGENY AND ITS ROLE IN MOLECULAR BIOLOGY

Our aim is to underline, throughout a complex history which lasted almost half-a-century, the development of interpretations. In that respect lysogeny "occupies a priviledged position at the cross-roads of normal and pathological heredity, of genes and viruses," writes André Lwoff in his fine critical comment (A. Lwoff, "Lysogeny" <u>Bact. Rev.</u>, 1953, 17:270). There were divergent interpretations as soon as Twort (1915) and d'Herelle (1917) discovered the bacteriophage. Examples are cited.

We deliberately give a place of importance to the work of Eugène and Elisabeth Wollman at the Institut Pasteur in Paris. From 1920 to 1943 the historian of science can observe how they built up the interpretation of the nature of bacteriophage and of lysogeny in terms of Mendelian heredity. Thus the concept of "paraheredity" for the transmission of characters through the external environment. Thus the linkage gene-virus, and even more the linkage heredity-infection, until then regarded as incompatible and today identified through lysogeny. Lysogeny was first rejected by the Delbrück group. But thanks to <u>integrated</u> inquiries the question of lysogeny was picked up again and definitely elucidated at the Institut Pasteur from 1949 onwards under A. Lwoff's direction and inspiration.

What do we learn from knowing the mechanisms of lysogeny ? First the notions of "prophage" and "temperate phage". Then the elucidation of the genetic determinism of lysogeny marks the triumph of genetic analysis. It combines with the clarification of the genetic determinism of bacterial sexuality. Thanks to high frequency recombination discovered by Cavalli and Hayes and to the characteristic of crossings between donor and recipient bacteria, lysogenic and non-lysogenic, prophage find their specific location on the bacterial chromosome. Hereafter the mechanism of genetic transfer can be studied. We then show the importance of zygotic induction and that of the linkage between mechanisms of bacterial immunity and those of repressive control of enzyme biosynthesis. We conclude with the notion of episome. Thus comes into play the work of F. Jacob, J. Monod and Elie Wollman. By saying that "position is the fourth dimension of prophage", A. Lwoff showed, through one of its aspects, that here was indeed raised one of the fundamental problems of molecular biology.

Lily E. Kay

Graduate Student, The Johns Hopkins University, Baltimore, U.S.A.

THE TECHNOLOGY OF KNOWLEDGE: THE TISELIUS APPARATUS AND BIOLOGICAL RESEARCH IN AMERICA, 1938-1948

In the numerous studies in the history of molecular biology there is little discussion on the importance of laboratory apparatus and experimental techniques. As a result, the process by which biology evolved into a highly sophisticated laboratory enterprise has been largely ignored. This paper aims at filling part of this gap by examining one of the most important technical advances in molecular biology, the Tiselius apparatus and its relation to biological knowledge.

The paper recounts the introduction, in the late 1930s, of the electrophoresis method by Arne Tiselius, which revolutionized the course of biological research. That method, by seperating molecules on the basis of their charge, made possible not only the isolation of single proteins from complex natural mixtures, but also became an indispensable tool for studying the behavior of proteins in solutions and at surfaces.

Only a handful of laboratories, mostly those which were supported by the Rockefeller Foundation, were fortunate enough to possess the expensive Tiselius apparatus, and as a result these laboratories accumulated information that quickly placed them in the vanguard of knowledge. As the paper shows, electrophoresis became a focus for the organization of inter-disciplinary research, which guided experimentation in several branches of molecular biology.

Terry Stokes

Research Associate, Deakin University, Australia

The Role of Molecular Biology in an Immunological Institute

The Walter and Eliza Hall Institute of Medical Research in Melbourne is pre-eminent in Australian bio-medical science. Best known for its immunological work, it also has an international reputation in auto-immunity, oncology and, more recently, immunoparasitology and molecular biology. The Institute's most celebrated Director, the Nobel Laureate Sir Frank MacFarlane Burnet (1944-65), would have no truck with molecular biology; holding that, "overall, its human implications are sinister rather than promising [*Changing Patterns*, Heineman, 1968, 187]." But Burnet's successor, the current Director, Sir Gustav Nossal established a Molecular Biology Laboratory in 1971, which began by examining the messenger RNA of immunoglobulin. At that time there was a good deal of scepticism in the Australian scientific community about the value of molecular biology. Prevailing techniques were cumbersome, and meant an initial lack of productivity which made the position of the Molecular Biology Laboratory's somewhat precarious. It was, however, secured when, in the middle of the decade, the research at last began to yield significant findings. The focus of the work, assisted now by the powerful recombinant DNA (rDNA) techniques which were becoming available, turned to a fruitful investigation of the molecular genetics of the immunoglobulin heavy chain. During the later seventies, two senior scientists were added to the molecular biology group. But an Australian version of the Massachusetts affair developed when, in 1977, controversy erupted over the installation at the Hall Institute of Australia's first biological containment laboratory for rDNA research. By 1982, the Molecular Biology Laboratory had grown into the Molecular Biology Unit - one of eight at the Institute. Attention shifted to oncogenes, and culminated in elucidation of the operation of an important oncogene. At the same time, rDNA techniques also began to be applied to aid the search by the Immunoparasitology Unit for a vaccine against malaria. To pursue this, one of the senior scientists from the Molecular Biology Unit moved to the Malaria Laboratory. Indeed, most Units at the Hall Institute are now clamoring for access to rDNA technology. The Hall Institute is currently grappling with the problem of satisfying this demand without undermining the integrity and autonomy of its Molecular Biology Unit.

BERNARDINO FANTINI

DIPARTIMENTO DI GENETICA E BIOLOGIA MOLECOLARE + UNIVERSITA' DI ROMA

THE ORIGIN OF THE OPERON MODEL

Abstract

The paper reconstructs the theoretical and experimental aspects implied in Jacob's and Monod's operon model of cellular regulation. This model was the result of an unforeseeable synthesis between two different research programmes, enzymatic induction and lysogeny. Both were developed at the Institut Pasteur in Paris, in continuous and close cooperation with the few scientific centers that in a small and tightly knit scientific community were constructing the new discipline of molecular biology.

Both biochemistry and genetics contributed to the theoretical structure of the regulatory genetics. The operon model was based mainly on purely genetic criteria arising from the phenomena of lysogeny, studied by François Jacob, and from the linkage between permease and galactosidase established by Monod. The synthesis between genetic data and concepts, and biochemical findings provided a way to generalize, by analogy, the model to different genetic systems. Thus, the interchange of concepts and results endowed the model with a high predictive value.

Margaret Somosi Saha

Scholar-in-Residence, University of Virginia

"ERWIN BAUR AND THE STRUGGLE TO PROMOTE PLANT BREEDING RESEARCH IN GERMANY"

The significant and often ground-breaking investigations of the German plant geneticist Erwin Baur (1875-1933) on such diverse topics as lethal genes, variegation and speciation rightfully placed him in the forefront of international genetics research in the early decades of this century. But even more remarkable was his unflagging ambition to apply the fruits of research from this new discipline towards practical ends. He, therefore, formulated a detailed program whereby the state, scientists and private industry would work harmoniously together to produce a scientifically-based planned economy rendering Germany economically and agriculturally self-sufficient as well as eugenically strong. Few episodes in the history of science can better illustrate the subtle interplay between scientific, intellectual and socio-political factors than the motivations behind Baur's tireless campaign to implement his vision, the formidable obstacles he encountered, and the eventual, albeit partial, success he enjoyed.

With his acceptance of a position at the Landwirtschaftliche Hochschule in 1911, Baur's penchant for practical applications developed into a passion as he embarked on his lifelong attempt to institutionalize his program for scientific plant breeding. He quickly encountered a number of obstacles: conservative government ministries, defensive private concerns, traditional agricultural workers, and most importantly, an academic community still dominated by a "mandarin" ideology and highly fearful of new disciplines, specialization and practical applications which might taint their pursuit of the pure <u>Wissenschaft</u> ideal. But Baur was no revolutionary. He conducted his campaign not by repudiating this ideal but by molding it to the scientific and socio-political conditions of the time. For example, although he endorsed the need for specialization and practical applications, he insisted that they be viewed in the context of a larger whole. Toward this end he devoted the last years of his life to formulating a comprehensive theory of evolution which synthesized his practical as well as theoretical work. He also argued (particularly after WWI) that technology be placed in the service of higher goals--the unity and cultural destiny of Germany. By adapting the <u>Wissenschaft</u> ideal to changing conditions Baur made his program palatable to a society which wished to enjoy the benefits of an industrial age while retaining many of the comforting values of the past.

3.6

Denis BUICAN

Professeur à l'Université de Paris X

L'ACCUEIL DU MENDELISME ET DU MORGANISME EN FRANCE

La génétique reposant sur les lois de Mendel et la théorie chromosomique de Morgan rencontrèrent en France une résistance inhabituelle même dans les cas de mutation de paradigme scientifique. En effet, la nouvelle science de l'hérédité trouva dans la voie de son développement l'obstacle du néo-lamarckisme qui - avec le dogme de l'hérédité de l'acquis - s'opposa à la relative invariance des facteurs héréditaires - montrée par le mendélisme - et au paradigme probabiliste impliqué par ses lois et par la signification de la mutation dans la théorie chromosomique.

Informé du mendélisme depuis sa redécouverte - grâce à la communication de Hugo de Vries à l'Académie de Paris et, un peu plus tard, par les recherches de Lucien Cuénot concernant l'extension de la valabilité des lois de Mendel dans le règne animal - le monde scientifique français accueillit avec un scepticisme injustifié la génétique classique. Cet accueil présenté dans notre dernier livre (Denis Buican, <u>Histoire de la génétique et de l'évolutionnisme en France</u>, Editions P.U.F. 1984) sera analysé dans le texte de notre communication dans ses implications multiples et à la lumière de nos dernières recherches.

Le développement de la génétique en France présente un caractère exemplaire pour l'histoire, la philosophie et la politologie des sciences car, devant de nouvelles découvertes qui amènent une mutation de paradigme, se trouvent coalisées spontanément des structures universitaires rigides et conservatrices, des idées et des idéologies périmées et des personnalités scientifiques d'arrière-garde. La communication brosse le tableau de ces imbrications et implications multipolaires en montrant au passage les constructions arbitraires élevées par le néo-lamarckisme tardif et hyper-tardif pour couvrir un espace scientifique naturellement dévolu à la génétique classique.

En s'arrêtant à la création de la première chaire de génétique en Sorbonne (1945) - date après laquelle la génétique trouve la voie d'un développement normal - la communication met en évidence une série de repères fondamentaux concernant le développement de la biologie en France, en apportant ainsi des données historiques indispensables à une réflexion scientifique et philosophique.

FISCHER Jean-Louis

Biologiste (Doct. Hist. Sci.),CNRS (Centre d'histoire des sciences
Université Paris I).

La IV e conférence internationale de génétique (Paris 1911)

Ce n'est pas par hasard si la IV e conférence internationale de génétique s'est tenue à Paris en 1911; car, en effet, si la science française n'était pas alors particulièrement favorable à l'aspect théorique de la génétique naissante, elle était en revanche particulièrement active en ce qui concernait l'aspect pratique de cette discipline (par exemple, l'hybridation végétale pour l'amélioration des races et variétés végétales). Aussi c'est à la Société nationale d'horticulture de France que fut confié le patronage de cette IV e conférence, dont l'un des animateurs fut Ph. de Vilmorin. (on notera aussi la présence de quelques grands noms qui ont illustré l'histoire de la génétique comme, E. von Tschemak, W. Bateson, W. Johannssen, A. Delcourt et E. Guyénot etc.).

Il n'est pas inintéressant, non plus, de souligner que le bureau effectif de cette IV e conférence est présidé par le néo-lamarckien Y. Delage qui préside également trois séances de travail sur cinq. Nous présentons une analyse du contenu théorique et pratique de cette IV e conférence vis à vis de l'esprit scientifique français alors en vigueur.

Barbara A. Kimmelman, University of Pennsylvania, Philadelphia, Pa.

Graduate Student, History and Sociology of Science, University of Pennsylvania

Agricultural Breeders and Mendelism: A Crucial Alliance for American Genetics

Between the rediscovery in 1900 of Mendel's paper on intravarietal hybridization in plants and the first publications by T.H. Morgan and his Columbia University graduate students c. 1900-15, agricultural scientists interested in the phenomena of inheritance literally transformed themselves from scientific practical breeders into geneticists. They were aided in this transformation by various resources, intellectual, cultural, and financial, directly referable to their affiliation with agricultural institutions. These included, generally, the injection of science into the agricultural curriculum; increased support of agricultural research in the form of the Hatch Act of 1887 and the Adams Act of 1906; and the realization of the professional ambitions of scientists who sought and found research opportunities within the state agricultural college/experiment station complexes.

More specifically, experimental plant breeders at agricultural institutions, who had contributed to the late-ninteenth century crescendo of interest in hybridization as both a breeding technique and as a powerful tool in the experimental study of evolution, were exposed to a passionate advocacy of Mendel by William Bateson and others at the Second International Conference on Plant Breeding and Hybridization at New York City in 1902. Their remarkably positive response reflected their intellectual interest in inheritance and the urgency of the practical demands placed upon them by their agricultural constituency. Mendel appeared to offer them both an explanatory model and a predictive, interventionist technique which could revolutionize plant improvement. Agricultural breeders pursued their interest in Mendelism within various agricultural institutions and associations, and founded new organizations devoted to research in breeding and Mendelism. Most important from the standpoint of disciplinary development, breeders established at agricultural colleges and stations across the country courses and programs in experimental genetics. At three, Cornell, Wisconsin, and Berkeley, independent departments in genetics were established between 1907 and 1913, devoted to basic research, graduate training, and ultimately undergraduate instruction as well. The extent and depth of these and parallel, if less fully realized, developments at agricultural colleges and experiment stations provided the broadest possible institutional base for academic genetics between 1900 and 1915, and in my view explains much of the remarkable success of U.S. investigators in genetics in the Progressive Era.

N. Roll-Hansen
University of Oslo

Swedish plant breeding: The Svalöf institute

Svalöf was founded in 1886 with the practical purpose of improving the seed stock of farmers in Southern Sweden. Within a few years the new institution had pioneered a system of plant breeding that became a world-wide model. Foreign visitors were impressed by the efficient production of new valuable varietes as well as by the system for introducing them to the farmers. The work at Svalöf is an excellent illustration of the close interaction between practical plant breeding and theoretical genetics around the turn of the century. The practical success was built on advances in biological theory.

The break with orthodox Darwinism in the form of mass selection, the "German method" of plant breeding, was crucial to the successes. This method consisted in selecting a number of plants with a desired character and planting their seed together as one mass. At Svalöf the results were disappointing. The progeny was inhomogeneous, and in many cases the selection did not affect inheritance to any appreciable extent. In 1893 Nils Hjalmar Nilsson therefore introduced pedigree-culture. He selected <u>individual plants</u> and grew their descendants separately. This was an anticipation of the principle of pure lines which Wilhelm Johannsen introduced a decade later. Hugo de Vries also came to see the results at Svalöf as a most important confirmation of his mutation theory.

Hybridization was developed into an effective breeding method by Herman Nilsson-Ehle during the first decade of this century. A visit by Erich Tschermak in 1901 was an important source of inspiration. Nilsson, backed by de Vries, was sceptical, and thought selection of mutations was the best procedure. Like Nilsson's innovation in the preceding decade Nilsson-Ehle's new breeding method was built on progress in genetic theory. In three publications between 1908 and 1911 Nilsson-Ehle was the first to demonstrate how inheritance of quantitative character is explained by multiple Mendelian factors.

This paper will attempt to trace the development of the theoretical basis of the breeding work at Svalöf from orthodox Darwinism, through the pedigree and mutation method of Nilsson and de Vries, to the modern cross-breeding of Nilsson-Ehle.

Joan Mason

Faculty of Science, The Open University, Milton Keynes MK7 6AA, UK.

CHANGING ATTITUDES IN SCIENCE EDUCATION FOR WOMEN IN GREAT BRITAIN;
THE OPEN UNIVERSITY EXPERIENCE

This century, particularly its second half, has seen complex changes in the education of women in science, mathematics, and technology. Although female participation has improved overall, some retrogression is evident. Less girls opt for science when single-sex schools have become mixed (unless special efforts are made) and most of the senior positions in former women's colleges have gone to men. The recession, cuts in Government spending, and merging of institutions, all hinder women's progress.

The title of this paper contains a deliberate ambiguity. Since its inception in the late 1960s the Open University has been conscious of its role in social change through adult education, particularly of disadvantaged groups. It provides multi-media distance education (with BBC TV and radio, Summer Schools, etc.) and requires no previous qualifications. The flexibility of part-time, home-based study is peculiarly helpful to women.

The O.U. now has 67,000 undergraduate, 850 postgraduate, and 27,000 associate students (plus 69,000 graduates). Across its history, significant changes are documented. Only 27% of the first students were women (resembling the proportion in conventional universities, though not their new intake, then a third). Nearly half of arts students were female, a third in social science, only 13% in science, 11% in maths., and 5% in technology. Now, with encouragement, we have near-parity of the sexes overall, and a greater proportional increase on the science side: on foundation courses, a third in science, a quarter in maths., and a fifth in technology; in science as a whole, a higher percentage of women than in the conventional universities. The pattern is familiar. Women form over half the population in the 'human' sciences (biology, education, food resources): falling to a third for chemistry, earth science, pure maths., statistics, and history of science, to a quarter for computing, a fifth or less in physics and applied maths., and less than a twentieth in engineering or electronics.

Affirmative strategies are still being developed, helped by WISE '84, the Women into Science and Engineering year promoted by the Engineering Council and the Equal Opportunities Commission. Some Government funding has been obtained for an O.U. Women in Technology project now in its fourth year, to help women to return to engineering, science, or mathematics after the family break.

Remarkably, in all the O.U.'s history and in all faculties women have done better than men (except in engineering, electronics, and more advanced physical science). Overall, the increment in pass rate for women compared to men <u>increases</u> in the sequence: social science, educational studies, arts, science, technology, mathematics. There are grounds for hope that deficiencies in the traditional education of women are being remedied, in the University of the second chance.

Ann B. Shteir

Associate Professor of Humanities, York University, Toronto, Cana[da]

FROM WAKEFIELD TO BECKER: INTRODUCTORY BOTANY BOOKS AND WOMEN

Botany figured in formal and informal education in England from the late 18th century on, and many books were written to introduce students to a fashionable and emergent science. Women were often singled out as the audience, and were themselves also prominent as popularizing writers. Priscilla Wakefield's <u>An Introduction to Botany</u> (1796; 11th ed., 1841) shows the format of early female-specific texts: familial, narrative, and aiming to teach women wider lessons about life through the study of plants. During the same period the standardized botany text takes shape: depersonalized, decontextualized, and ostensibly sex-neutral; and Lydia E. Becker uses the model in her <u>Botany for Novices</u> (1864). But is that the only way to write introductory botany books for women (and in general)? The changing format, partly due to directions in 19th-century British botany, is a reflection of late 19th-century feminist thinking about the locus and purposes of science education for girls and women.

Peter Dilg

Institut für Geschichte der Pharmazie, Marburg, FRG

Naturwissenschaftliche Literatur für Frauen im 18. und 19. Jahrhundert

Während die Beschäftigung mit dem Themenkomplex "Frau und Wissenschaft" ständig zunimmt und in diesem Rahmen hauptsächlich Aspekte wie "Frauen in der Wissenschaft" oder "Frauen als Wissenschaftlerinnen" berücksichtigt werden, scheint bislang die naturwissenschaftliche Literatur, die namentlich im 18. und in der ersten Hälfte des 19. Jahrhunderts speziell für Frauen geschrieben worden ist, weniger Beachtung gefunden zu haben. Es soll deshalb versucht werden, zumindest einen kursorischen Überblick über diese spezifische Gattung der "Frauenliteratur" zu vermitteln, d.h. einige dieser (fast ausschließlich von Männern verfaßten) Werke näher vorzustellen, nach ihrer jeweiligen Zielsetzung, ihrer Stoffauswahl und ihrer literarischen Form (Briefe usw.) zu analysieren und im Kontext der sonstigen, also für ein männliches Publikum geschriebenen Fachliteratur vergleichend, d.h. mit Blick auf eine mögliche Geschlechtsspezifität, zu werten. Dabei bleibt das medizinische Schrifttum zur Entbindungs- oder Hebammenkunst ebenso außer Acht wie etwa dasjenige zur Kosmetik, da sich dieses auch schon in früherer Zeit naturgemäß an die Frauen richtete; vielmehr gilt das Augenmerk einzelnen Werken vornehmlich über Botanik (z.B. von Rousseau, Batsch, Lindley), allgemeine Naturgeschichte (z.B. von Unzer, Suckow, Bischof) sowie Chemie (z.B. von Meurdrac, Hochheimer, Geitner, Lampadius, Marcet, Holger) und damit den ersten Ansätzen, die naturwissenschaftliche Erziehung und Bildung des weiblichen Geschlechts zu fördern.

Sally Gregory Kohlstedt, Syracuse University, Syracuse, New York, U.S.A

SCIENCE OR SENTIMENT? WOMEN AS NATURE-STUDY TEACHERS, 1890-1920

Nature-Study curriculum was developed to introduce phenomenon and ideas about science to school children. It was implemented in both urban and rural settings in the United States and elsewhere, encouraged by faculty at teachers colleges and at universities with large education departments such as Columbia and Chicago. Women, who predominated in the educational work force, established programs to suit local needs as well as meet broader objectives. The most competent and enthusiastic wrote textbooks, supervised city-wide programs, and became officers in the professionally-oriented Nature Studies Association.

Their work had several key components. Advocates were committed to education which coincided with current theories of educational philosophy and psychology. They insisted on placing scientific knowledge into a broad ecological framework, often using such techniques as school gardens and nature walks. They also used a variety of approaches, from literature to technical apparatus, in order to interest their students in the natural environment. Their typical reliance on a deductive approach to inquiry and their insistence on integration rather than specialization contributed to reaction among university scientists. The challenges to nature study were usually vague and provided no alternative curriculum. Compounding the debate about the value and effectiveness of nature study was the fact that the teachers were usually women and the challengers inevitably men, so that gender assumptions and assertions were a not-so-hidden aspect of the confrontation. Were women, some wondered aloud, willing and able to teach science in ways which trained young people for advanced work?

The debates over nature study highlight the unsettled state of the practice of science and of public conceptions of science at the turn of the century. Institutional and ideological factors played a significant role in these struggles and the outcome strengthened, at least on the surface, the authority of university-based research scientists and educational experts.

Dr. Joy Harvey

Assistant Professor, Sarah Lawrence College

Scientific Medicine and Radical Politics: American women studying medicine in Europe from the 1860's to the 1890's.

    A careful analysis of the American women who sought to study medicine in Europe in the mid to late nineteenth century reveals patterns ranging from a single year's study in order to learn a new technique to the obtaining of formal degrees from Zurich and Paris. Along with the exposure to the new scientific medicine came , for some of these women, an exposure to radical republican politics in an era of change. An outstanding example of a woman who wrote about both of these experiences was Mary Putnam-Jacobi whose ground-breaking presence at the Ecole de Medecine in Paris during the last years of the Second Empire through the Paris Siege and the Paris Commune resulted in her radicalization. Upon her return home, she , like her co-student , the Englishwoman, Elisabeth Garrett Anderson, became an outstanding proponent of women's rights as well as a model teacher of experimental medicine. The patterns of women later studying in Germany, Switzerland and France are illuminated by comparison with her example.

Ann Hibner Koblitz

Member, School of Social Science, The Institute for Advanced Study, Princeton, NJ USA

Russian Women at Zurich University in the 1860s and 1870s

In the 1860s, young women of the Russian intelligentsia and nobility were encouraged by the social philosophy of the nihilists to seek education and careers in the natural sciences and medicine. When opportunities in their native country closed to them, the most determined, committed, and advanced of the nihilist women went abroad to study in Western Europe. The majority of them went to the university and polytechnical institute in Zurich. Statistics for Zurich University show that a total of 203 women were enrolled as auditors or students between winter 1864-1865 and summer 1872. Of these, there were 23 English, 10 Swiss, 10 Germans, 6 Austrians, 6 Americans-- and 148 Russians. The Russian women overwhelmingly chose to major in the natural sciences and medicine.

This paper will concentrate on the Russian women's experiences in Zurich before the tsarist government disbanded the women's student colony in 1873. Attention will be paid to their stuggles to gain admission to Swiss institutions of higher education, and the reactions of Swiss society to the influx of nihilist women. In addition, the achievements of this first generation of Russian women in scientific research and clinical practice will be noted.

Liisa Hayrynen

University of Joensuu; Academic Careers Study 1965-82

Women in the changing medical profession - A longitudinal case of 34 female doctors

This paper concerns with the life-course of 34 woman medical doctors whose vocational interests and backgrounds were investigated, in 1965, in conjunction with the entrance examination of the Medical Faculty of Helsinki University; they participated in 1982 in a follow-up study together with several other groups of earlier university applicants. The comparative group consists of 74 male medical doctors of the same cohort.

In 1965, about 70 percent of the applicants in medicine were males. Student radicalism did not touch Medicine as it touched the fields of Social Sciences, Architecture, or the Humanities. In the 1970s, large changes in the medical institution included foundation of a net of public health districts and municipal health centers. This may have increased bureaucratization of medical work; in 1982 a general strike of medical doctors in public hospitals, indicated a change in the earlier upper-class role.

The majority of these 34 women came from the upper or middle social stratum. 60 percent of them indicated, in 1965, Medicine as the most favoured profession. They emphasized altruism and people as the vocational goal; male entrants more often appreciated a secure or an independent career. Female medical students generally reported a sharper growth of vocational self-determination during the university years, then their gender fellows in the liberal faculties. In 1982, about half of the earlier medical students described their university climate in positive expressive terms; still, there were almost a similar amount who felt that the earlier study atmosphere had been repressive.

Many of these 34 females experienced a stabilization of the professional status in the early 1980s, though a large group reported a steadily progressing development. Still, their general life emphasis is more on the family circle or the local sphere; the identification with the professional community is somewhat less prominent. A small proportion has been able to implement their research orientations. There are also some indications of shifts in priorities and goals.

Brigitte Hoppe

Professor, Institut f.Gesch.Naturw., Univers. München FRG

WOMEN'S WORK IN NATURAL AND APPLIED SCIENCES IN EARLY MODERN GERMANY — DILETTANTISM OR SCIENCE?

In former Germany no official institutions for education of girls were established and girls were not allowed to attend colleges and universities till the beginning of our century. Nevertheless since the early modern times many female persons excelled in different fields of natural and applied sciences. But only a few main works were mentioned in the historical literature. Therefore we will look also on a great deal of smaller and mostly unknown works of women and ask for their scientific value and historical signification.

The paper relies on a review of original published and inedited works on dietetics, pharmacy, alchemy, chemistry, botany, zoology and astronomy of female authors from the 17th to the 19th centuries. The activities on different levels of empirical research, as assistant work, collecting of natural objects and observations or independent new scientific research are explained and compared with similar contemporary scientific works of male scientists. Further we ask for the importance of the results of such works of female scientists for the development of natural sciences in earlier times: were those works negligible as products of dilettantism or had they an effect on the rise of modern science? The answer on these questions will elucidate the social background and conditions of the development of natural sciences.

Judit BRODY

Curator, Science Museum Library, London, U.K.

PATTERN OF PATENTS: EARLY BRITISH INVENTIONS BY WOMEN.

The inventive powers of particular socio-economic groups are not accurately reflected in the patent literature: thus prior to 1852 the cumbersome and expensive official procedures affected the number of patents taken out by women, while in the latter half of the 19th century social pressures served the same function. Patents of invention granted to women in Britain gradually decreased as a proportion of all patents from 1.6% to 0.4% between the years 1650/1700 and 1800/1850. While over 50% relate to the traditionally female spheres of home and wearing apparel, 25% are technical innovations outside the home and about 15% can be described as chemical or pharmaceutical - the last figure being approximately the same as for male patentees. The inventions range from spectacular failures such as Lady Vavasour's cultivating machine which was found to compact the soil instead of aerating it, to successes like Isabelle Lovi's philosophical beads which were not only patented but also manufactured and sold by her and were highly praised by David Brewster. The solid core of the patentees were spinsters and widows; many were themselves in business as corsetiers, milliners or instrument makers.

Dr. Daryl M. HAFTER

Professor of History   Eastern Michigan University

Women's Use of Technology in 18th c. Rouenaise Guilds

The persistence of women in certain guilds of 18th century Rouen gives historians an example of female entrepreneurial activity that has high status and technological sophistication rare in the period. While women had been enrolled among the founding members of many French guilds in the 14th century, they had been largely kept out of guilds by the 16th century. In contrast to the majority of female workers with their minimum pay and auxiliary status, women who remained in guilds became masters in their own right and enjoyed the powers of their privileged status. The Rouen guilds provide us with a case study of the effect of guild membership which had the power to lift the disability intendant upon women from their economic life.

In the ten guilds with female or mixed gender membership--the spinners, hosiers, embroiderers, old linen drapers, new linen drapers, fashionable plume makers, religious vestment seamstresses, makers of passementerie, drygoods merchants, and tailors-- women participated fully in the activities of the guild. They became guild officers, enforcing regulations and apprehending those who worked or sold goods without guild membership. Sophisticated in the use of the legal apparatus of the old regime, they defended their members by hiring lawyers, testifying in court, and petitioning the king's officers.

These craftwomen were in full command of their technology. They trained apprentices and judged masterpieces. They deliberated on the use of new technology and defended their members from encroachment by other guilds. It was their license to perform the technical operations that gave them status within their families and the legal right to manipulate institutions in the world at large for their own benefit.

Elvira Scheich, Technische Universität Berlin, Germany

WHAT KEEPS THE WORLD GOING ?
A FEMINIST COMMENT ON THE HISTORY OF THE IMPETUS-THEORY

This theory of motion prepared the foundations of Classical Mechanics. At the same time it was a theory of economics, with it were created the modern conceptions of money, labour, and property, which led to the economical theory of the 18th century.

It is mainly the economical part, which gives the chance to ask: what do the parts of theory taken as evident tell about women? In a very indirect and implicit form they exclude women from activities in science.

The new ideas brought up by the impetus-theory - in physics as well as in economy - originated in trade and industry of the growing medieval cities. Early capitalistic forms of enterprise effected the transformations in social economy and economical attitudes towards nature at the end of the Middle Ages in many aspects. The denial of reproductive labour in the theory of money, which could explain its function as capital, corresponded to the (technical) exploitation of natural ressources.

The range of objectification was in the 14th century limited by the traditional unity of reproduction and production in a rural subsistence economy and by the corresponding social realtions and ethics. Looking at the history of female work and labour the whole depth of the conversions required by the proto-capitalistic industry can be shown.

The realities of womens life and the image of feminity, that had been created by the medieval society were not yet tied to motherhood, housework and sensibility. But it was the general discrimination of women that provided the starting points for the theoretical objectifications by the impetus-theoreticans.

The historically following view of reproduction (of human labour) as a purely natural ressource completes the conception of production, which determined the economical and physical statements of the impetus-theory, and led beyond the limits which confined the thinking of the 14th century.

MADHURI SHETH
ASSISTANT PROFESSOR
NATIONAL INSTITUTE FOR TRAINING IN INDUSTRIAL ENGINEERING, BOMBAY

"WOMEN'S CONTRIBUTION TO TECHNOLOGICAL INNOVATION : PRODUCT AND PROCESS"

History of human civilization is essentially the history of innovations in technology. Unfortunately, technological innovation has come to be identified mainly with men's work involving production of goods and services. In the process, men's relationships with others with whom they have to live, and those who are affected by such increasing capacities have been left out. Women are the largest group so affected.

The imbalances in human relationships which are traced to technological developments are now shifting our attention from products of technology to processes of technology. Looking into the processes of technology, it is also being slowly realized that women have been responsible for some fundamental technological innovations. Very few societies have so far thought it fit to take cognizance of the contribution of women. This awareness is now leading to innovations in organizational technologies. It is in fact in this area where women's contribution to technological innovations seems to be most significant in terms of its impact on the individual and the society.

These organizational innovations result in, for example, the creation of an organisation or an institution which explores, diagnoses and studies local needs and local solutions to these needs, and then links these two with technological change. The technique involved is the intimate knowledge of the critical issues facing the target/problem group of persons, say women, and a modest attempt either to improve productivity or to reduce time and energy spent on the task, or to retrain women from one technology to another.

Thus when women's organizations take technological decisions the welfare of women is not a byproduct of technology such as a creche put up by a factory. Technology in their decisions becomes a compliment to women's welfare. The major objective of these organizational innovations is to make not only women but also communities economically self sufficient and self reliant.

Dorothy K. Stein

none

Introducing Ada: Myth and Symbol in Computer History

In the last few years, almost anyone with an interest in computer history, and large numbers of people without, have heard that Byron's daughter invented computer programming. This is despite abundant evidence, freely available at least to scholars, that her abilities as a mathematician and achievements as a computer programmer were in fact illusory. Ada's reputation has been based on statements by herself, by Charles Babbage, the originator of the idea of the general purpose computer, and on her authorship of a paper describing his projected machine. As I have shown previously (Stein, Victorian Studies, Autumn, 1984) she herself was well aware of her own technical limitations. Babbage's testimonial was self-serving. with the object of promoting his invention, and was in any case written twenty years after the event. Moreover, a careful reading of the paper itself shows that, far from being a clear and masterly exposition of the structure and logic of the computer, it was a rather mystical tract that dwelt on the inventor's mechanistic views of theology and the workings of capitalist economics. In addition, by means of a subterfuge, Lady Lovelace repeatedly made claims for the ability of the planned engine to perform symbolic computations, which it did not in fact possess.

Why then has there been such a growth in Ada Lovelace's reputation, what purposes does such a myth serve, and for whom? In renaming after her a computer language commissioned and developed for the needs of the Department of Defense, the military establishment reenacted a medieval chivalrous ritual, in which preparations for war, death, and destruction were dressed up in romantic trappings. Computer scientists were able to give their rather hard, abstruse and daunting craft a more human and attractive cover, and, above all, disguise the extreme gender imbalance of their profession. The public at large, scientific or otherwise, welcomes sex in any form.

The dangers for feminists, who are busy trying to promote Ada as a neglected scientific genius are most insidious. Not only do they beg the question of whether women (unlike men) must be geniuses to pursue a career in science, but, when her reputation is exposed, they risk, once more, having to contend with the assumption that because a particular woman has been shown to have little talent for mathematics, it follows that all do. The issues and implications of the relationship between women and science or technology--including Ada's own--are far deeper and more interesting than the search for heroines, genius, or even success, will admit.

Dr. Aihwa Ong

Assistant Professor of Anthropology, U.C., Berkeley

"THE PRODUCTION OF SUBJECTS IN HIGH-TECH INDUSTRIES"

This paper discusses the encounter between third world women and high-tech industries as a cultural process. Techno-scientific knowledge produces a technology of politics in new industrializing milieux. Power embodied in electronics procedures is all the more effective in its ability to conceal mechanisms of domination and control. It is argued that high-tech production processes located in the third world, simultaneous with the manufacturing of micro-components, reconstitute gender hierarchy and produce a new sexuality among working class women.

High-tech industries are characterized by the dispersal of research, marketing and production systems and extreme fragmentation of the labor process in time and space. Such procedures enhance individuation and mobility among workers treated as interchangeable commodities (or organs: eyes, hands). Simultaneous with such manipulation of labor, corporate culture fosters, through company beliefs and practices, a multiplicity of interchangeable gender images. In West Malaysia, for instance, the creation of a new sexuality is a dominant effect of the scheme of disciplinary techniques in electronics companies which employ a large female labor force. The management of their emotions, however well-intentioned, produces empowering knowledge which enhances and disguises the subjection of labor to capital. Sexuality becomes a set of practices for the reconstitution of subjects, a field of meanings to be deciphered, and an economy of discourses which complement the operation of power in daily production relations.

Creativity thus occurs in the cultivation of self, both as a mode of acquiescence, and of resistance to shifting domains of control in everyday life. This fashioning of new identities results in a local cultural impoverishment but facilitates survival and the reconstruction of new connections in the spatial, social and cultural dislocations of working populations in late industrial capitalism.

Depending on the possibilities of conducting research, a comparative perspective will be provided by discussing management techniques utilized in electronics companies located in the Silicon Valley.

Else Høyrup

Roskilde University Library, Roskilde, Denmark

WOMEN MATHEMATICIANS

Women do not very often become scientists. But there have been quite a few women mathematicians. I shall talk about 4 of them.
When making a <u>collective biography</u> of women scientists, I propose a <u>time-structure</u>: The first period (the period of amateurs) before the admission of women to the universities. As a representative of this period, I choose Mary Somerville (1780-1872). She was a Scottish-English scientific expositer and mathematician. The second period is the time of admission of women to the universities in the latter half of the 19th Century, when women could get an education, but rarely a job. As a representative here, I choose Sofia (Sonia) Kovalevskaia (1850-91). She was a Russian-German-Swedish expert in mathematical analysis. She did in fact get a job, - because she became a widow, which was honorable. And the third period, after admission of women to the universities, when women slowly began to be able to make a career, but rarely a regular one. For my purpose, I divide this period in a first one before about 1960 and a second one from the sixties onward. From the first of the periods of the 20th Century, I choose Käte Fenchel (1905-83). She was a German-Jewish-Danish group theoretician, who did not get a regular career, but was married to a professor of mathematics. From the last period, I choose Gerd Grubb (1939-). She is a Danish expert in partial differential equations, who has got a regular career, but who knows the problems of combining children with a career. There is also a dividing line around the Second World War, but I do not use this in my short presentation.

Caroline Claire Scanlan Murphy

PhD student Dept. of History of Science & Technology UMIST England UK.

THE EARLY HISTORY OF WOMEN IN BRITISH RADIOTHERAPY. C.1899 - 1939

Marie Curie brought the idea of radium therapy into the public eye at the Royal Institution in 1903, but already British women were involved in the development of radiotherapy. The international shortage of radium meant that this work was largely in the field of X-ray (or Roentgen) therapy until the Great War. During the twenties women were regularly employed as assistant radiologists doing important work in the establishment of effective radiotherapies. By the early thirties radiotherapy involving the use of both X-rays and and radium was becoming established as a separate specialty and women played an increasingly important role as medics, hospital physicists and laboratory research workers.

The Roentgen Society(1897), the first radiological society, admitted women to its broadly-based membership(both medical and lay) from its foundation. The exclusively medical British Electrotherapeutical Society(1902) did not accept women, with the result that lady radiologists joined the Roentgen Society. The exclusion of women from the London medical establishment resulted in women forming their own medical school and exclusively women's hospitals. In these hospitals women had to take on all the medical specialities, including radiology (both therapeutic and diagnostic).

The first major article on X-ray therapy published in the Archives of the Roentgen Ray (the earliest radiological journal) was by Margaret Sharpe LRCP LRCS in 1899. The most remarkable aspect of the work of this pioneer was the fact that on discovering in 1901 that as "...X-ray cure proper...is probably not due to any kind of a light ray but to electric currents.", she went back into print admitting that this was "...a flat contradiction of the statement [she] made in 1899.", a brave admission for any author to make: it was to be several years before her apparently contradictory explanations could be reconciled.

The universal lack of understanding of all radiations and their dangers resulted in large numbers of injuries to both patients and practitioners. This meant that few young men were attracted to radiology and that some positions were open to women. These positions were not only medical as at the same time there was an expansion in laboratory research posts. Several of the most important figures in early radiobiology (who provided important understanding of the effects of radiation on cells, tissues and whole animals) were women, the most famous of whom were Helen Chambers and Honor Fell.

In this paper I hope to show that the physicist Marie Curie (although the most famous), was not the only woman involved in the establishment of radiotherapy: others shared with her the dangers unique to radiation, as well as the more commonplace problems of research funding and career development.

Éva Katalin Vámos
research-worker,
Hungarian Museum for Science and Technology, Budapest

WOMEN AND SCIENTIFIC RECOGNITION

Though we find more and more women employed in research institutes and at universities these days, we find very few who received high recognition in their own field of science. Turning over the pages of all the 14 volumes of the Dictionary of Scientific Biography we find only 17 biographies of women scientists. Most of these women lived in the 19th and 20th c., few in the classical antiquity. It is usual to mention that in the 18th century women played an important role in science though not as researchers but as hostesses of the French type of "salons". But it was only from the very end of the 19th century that women could study at numerous universities /in Hungary from 1895 on/. Studies of some branches of science and engineering became possible for women only after World War II. That is why recognition for women in their own field of research will probably be part of the history of science of the 20th and 21st centuries. However, while access to intense research became possible for women in our century, many new and old questions are raised regarding the aims of women's studies and their careers in science. /A special history of women and scientific recognition in Hungary will also be given in the paper./

KONCZ, Katalin

Associate professor, University of Economics and Politics "Karl Marx" Budapest, Hungary

THE HISTORICAL PROCESS OF FEMINIZATION IN INTELLECTUAL PROFESSIONS IN HUNGARY FROM 1880 to 1980

The flow of women into intellectual professions is the product of the industrial revolution in Hungary, too. Their ratio is continually rising in the white-collarjobs: in 1880 it was 13 %, in 1910 20 %, in 1930 33 %, in 1960 45 % and 58 % in 1980.

The expansion of female employment is accompanied by two interrelated processes:
- women enter an ever larger field of jobs and are trying to find employment in careers formerly to be typically masculine;
- the number of jobs and occupations filled typically by women is expanding, and thus more and more become feminized.

Hungary may serve as a good example to show that the spreading of feminization has a clearly palpable historical course.In 1920 health, in 1949 education and culture became sectors employing primarily women. In 1980 all jobs in accounting and administration, two thirds in health, and half of them in education and culture employed women in a greater proportion. The ratio of women is the highest in all jobs at the lowest level of the hierachy.

The feminization of careers is accompanied by the devaluation of these jobs together with a counter-selection process.As result, the sectors employing a larger proportion of women offer less favourable than average income relations, and their prestige is diminishing. The reserve in also true: the feminization always begins in jobs and in a period where and when their prestige is declining. Ultimately, the cause and effect are interwoven in the self-maintaining and selv-reproducing mechanism, and it can be proved that the deterioration - also noticeable in the long run - of the quality of the labour stock in occupations offering less favourable conditions and losing their prestige begins to make itself felt.

Shu-ping Yao

Associate Researcher, Office of Science Policy Research,
Chinese Academy of Science, China

Women Scientists in China in the Last 35 Years

"Women Scientist in China in the Last 35 Years"
The article describes in brief the courseof Chinese women's rise to their present status of equality between the sexes from the lower stratum in the feudal society. It analyses the proportion and situation of Chinese women in different stages from the time of their studying in school to that of their entering the scientific circles, introducing emphatically the role they play in the scientific community and their achievements. Besides, problems which women scientists now confront, various views of pros and cons in the society concerning women's taking part in scientific research and prospects for the future are also mentioned.

David K. Allison

Historian of Navy Laboratories, Naval Material Command, Wash DC

THE EVOLUTION OF U. S. NAVY LABORATORIES, 1869-1984

Since World War II, the United States government has invested a substantial portion of its expenditures on scientific research and development in government laboratories. This has been especially true in the U. S. Navy, where in-house laboratories not only develop new and improved systems based on their own research, but also help the Navy evaluate technical developments made in universities and private industry.

While laboratories did not become a major part of the naval establishment until World War II, their history begins much earlier, with the establishment of the Naval Torpedo Station, Newport RI, in 1869. This paper will review the major phases of the evolution of the Navy laboratories from then to the present. It will explain what led the Navy to establish its major laboratories when and as it did, and how those laboratories have evolved. Establishing or expanding laboratories became a normal Navy response to major advances in science and technology. However the character of this response was always shaped by prevailing views. This institutional analysis, then, will reveal the changing perceptions Navy leaders have had of scientific research and development and its role in naval affairs. It will also display changing ideas about the most effective means for managing science and technology.

Paul Forman

National Museum of American History,

Smithsonian Institution, Washington, D.C. 20560

AMERICAN CONTEXT FOR QUANTUM ELECTRONICS, 1940-1960

While nuclear physics had already established its priority as a research field, and its orientation toward ever higher energies, before the Second World War -- so that the advent of nuclear weapons served mainly as a powerful booster -- quantum electronics owed its existence, as well as its rapid growth after 1945, to the wartime mobilization of physicists and engineers for work on radar, especially in the microwave range.

Prior to the early 1960s, when the mushrooming field of laser research and applications preempted the term, "quantum electronics" covered all those areas of pure and applied physics concerned with resonance phenomena in the interaction between electromagnetic waves and the quantized energy states of atoms, molecules, and condensed matter. Comprised therein were the fields of microwave spectroscopy, electron spin resonance, and nuclear magnetic resonance, as well as various double resonance techniques combining these and optical methods. Each and every was, on the one hand, inspired in some degree by the magnetic resonance method Rabi had so successfully applied to atomic beams just prior to the war, and, on the other hand, based more or less heavily upon wartime advances in high frequency radio electronics.

The physicists contributions to weaponry in World War II persuaded the US Army, Navy, and Air Force, as also the leading spokesmen for US science, that America's national security depended upon a "continuing working partnership" between the universities, industry, and the military, to prosecute a permanent technical revolution in armaments. This concept motivated essentially all of US government support for basic physical research in the postwar period, with more than 90% of the funds coming from the Department of Defense and the Atomic Energy Commission. Pegged at about 5% of the total US military expenditure for research and development, the funds rose steeply following US entrance into the Korean war and again in response to Sputnik. As President Eisenhower stressed on departing from office in January 1961, in consequence of the scale of this expenditure and the program it served, "the free university ...has experienced a revolution in the conduct of research."

That revolution, carried forward in the context of cold war, and, in its first decade, under the shadow of "the mania for secrecy" (L.V. Berkner), carried with it research agendas influenced by the attraction of military money, the instructions of military sponsors, and the fascination of military applications. Quantum electronics, standing at the center of this revolution, brought its most striking successes.

Alex Roland

Duke University

THE NATIONAL ADVISORY COMMITTEE FOR AERONAUTICS AS A MODEL OF GOVERNMENT-SPONSORED RESEARCH

The National Advisory Committee for Aeronautics (1915-1958) was for much of its history the premier aeronautical research organization in the United States; for many years it was arguably the most productive organization of its type in the world. Since its demise, many observers in the United States have recommended that it be revived in order to restore American aeronautical research to its former preeminence.

The NACA attributed its success in large measure to organizational arrangements. Among the essential features were independent status within the federal bureaucracy, direct access to the President, governance by an unpaid committee of government and civilian experts, coordination with other aeronautical research agencies, solicitous concern for the needs and interests of its clientele (primarily the military services and the aircraft manufacturing industry), scrupulous observance of the spheres of influence of other aeronautical research agencies, and avoidance of partisan politics. Some of these policies infact contributed to NACA success; others finally contributed to the agency's disappearance.

In retrospect, the NACA does provide a useful model for government-sponsored research. Its independent status, heavy reliance on committees, thorough-going coordination of research, and avoidance of politics all seem to have contributed significantly to the Committee's success. At the same time, many of these characteristics contributed to the NACA's demise and raise the question of whether such an organization can survive in the executive branch of the U.S. government. Features like concern for its clientele and observance of spheres of influence may have done more harm than good.

Robert Wayne Seidel

Research Historian, Laser History Project

FROM GLOW TO FLOW: A HISTORY OF MILITARY LASER R & D

The invention of the laser provided a new scientific and technological opportunity to the military planners. Like nuclear fission, coherent radiation suggested a variety of applications, many of which have been pursued over the past 25 years. The scientific and technological research programs of the military have been a function of the special properties of laser radiation, of the the special missions and of the institutional arrangements for research characteristic of the services. A comparison of these programs reveals the nature of the relationships between science, technology, industry ,and the military.
The discussion will focus upon three major laser technologies developed in the military R&D community: the solid-state, gas-dynamic, and chemical laser systems. Solid-state lasers were the first to offer the potential of scaling to high power levels, as well as the best candidates for range-finding and similar applications of laser technology. The introduction of aerodynamic technological concepts with the gasdynamic laser overcame many of the thermal problems associated with large ruby and Nd-glass laser systems, but introduced new problems which had to be solved at the engineering level because of attempts to rapidly scale the technology for military purposes by all three services. The chemical laser displaced the gasdynamic laser as the primary high power candidate in the 1970s despite its development into successful experimental devices because of peculiar advantages it offered in military environments.
Analysis of these developments indicates that many of the innovations in laser research have been stimulated by an intensive interest in lasers on the part of the military, that the military has been more than a purchaser of the technology, and has made significant contributions to the scientific understanding of the laser through its inhouse and contract research programs. The attempts to coerce laser technology in order to fulfill certain missions, however, have not always been successful, and, for much of its history, the laser has been a solution looking for a problem in the military. The role of its sponsors and promotors has, consequently, been of major importance in keeping laser R&D programs alive. The laser is the archtypcial product of modern military R&D.

John A. S. Pitts

Historian, Naval Research Laboratory, Washington, D.C.

UNDERWATER ACOUSTICS AND NATIONAL DEFENSE: A STUDY OF NAVY-INDUSTRY UNIVERSITY COLLABORATIVE R&D, 1950-1967

The U.S. Navy's involvement in collaborative R&D was a new development following Worl War II. Anticipating rapid advances in Soviet naval warfare technology, the Navy identified a need to accelerate fundamental research and exploratory development in fields critical to advanced Naval warfare systems. Toward this end, the Navy initiated a major effort to expand its R&D capabilities by sponsoring research in industry and universities. With little previous experience in collaborative R&D, the Navy entered into early relationships that varied significantly in scope and character.

This paper examines two collaborative arrangements that were radically different in nature, even though both were initiated in the same year, endured for approximately the same length of time, and were concerned with the investigation of a common problem. One was an arrangement that established Hudson Laboratories as a Navy contract laboratory administered by Columbia University. Under this arrangement, Hudson functioned as a "lead laboratory" for basic and applied research in physical oceanography and underwater acoustics. Over a period of 17 years, Hudson was the project management center for two large-scale projects in which 15 university, industrial, Navy and other Federal laboratories directly collaborated.

The second was an arrangement with Western Electric Company (WECO) and its research subsidiary, Bell Telephone Laboratories (BTL). Based on a one-page letter contract, WECO-BTL over a period of 17 years, investigated the phenomena of very low frequency sound in water and developed a large scale system for exploiting these phenomena. Thougout the course of this development, WECO-BTL, acting solely on the authority of the letter contract, handled virtually all of the R&D and rarely collaborated with industrial, university or Navy laboratories.

This paper reconstructs the histories of these two collaborative relationships, identifies the factors that made possible two such different relationships, and explains why both arrangements were finally abandoned.

Michelangelo De Maria, Università di Roma "La Sapienza", Rome, Italy

Arturo Russo, Università di Palermo, Palermo, Italy

COSMIC RAY PHYSICS IN THE USA DURING THE '30s: EARLY EXAMPLES OF COOPERATIVE RESEARCH.

In this talk we examine the role played by the Carnegie Institution of Washington (CIW) and the Carnegie Corporation of New York (CCNY) for the growth of cosmic ray physics, during the late '20s and early '30s, from a marginal sector of research to one of the most central fields of physics.

We focus our attention on the growth of scale in funding and the reshape of institutional and organizative patterns of cooperative research in this field.

In particular, we briefly discuss the approach of R.A. Millikan and his Cal Tech group to Cosmic Rays during the late '20s and the main features of his research program, also in the light of his early links with Officials of the CCNY and with military and political circles.

We then analyze A.H.Compton's conception of cooperative research, his links with the CIW and the main results of his 1932 World Survey of Cosmic Rays ("Latitude effect", "East - West effect", etc.), and reexamine the famous Millikan - Compton controversy, which took place at the AAAS Meeting of Atlantic City in December 1932, not simply in terms of a disagreement on the nature of primary cosmic rays (charged particles versus corpuscles), but as a competition between the two prime donne in order to get further support from the CIW and as a confrontation between two different approaches to cooperative research.

Under this respect, we finally discuss the scientific policy of the "Committee on Cosmic Rays" set up by the CIW, in december 1932, in order to cope with competing research programs in cosmic ray physics (Millikan et al., Compton et al., T.Johnson et al, etc.)

David H. DeVorkin

National Air and Space Museum, Smithsonian Institution

HOW SCIENCE WENT INTO SPACE, AND WHAT IT DID WHEN IT GOT THERE:
THE V2 ROCKET PANEL

In the summer of 1945, parts sufficient to reconstruct about 100 German V2 rockets were carried by the U.S. Army from Europe to the southwestern United States. The Army intended to test these rockets as quickly as possible to narrow the perceived missile gap created by Germany in the closing years of World War II. There was every indication that future wars were to be fought with atomic warhead-laden missiles, so all aspects of missile research became of immediate priority in the post-war era.

As the Army planned for rocket firings in late 1945, a number of individuals and small groups in the United States, in process of reorienting themselves to the post war world, turned to the possibility of conducting research on the nature of the upper atmosphere and near space from rocket-borne experiments. One major group, at the Naval Research Laboratory in Washington, D.C., evolved from wartime work on missile fire control and communications security, and another, at the Applied Physics Laboratory of the Johns Hopkins University, was created from a multiplicity of talents involved in the development of the proximity fuse.

In January 1946 the Army offered to these and other groups the use of warhead space in the captured V2 rockets, and from this interservice liaison, which also included the General Electric Company as the prime contractor in firing the V2 rockets at White Sands, a Rocket Panel was created that coordinated the scientific experiments to be flown. This panel, composed by direction of active participants in the design and construction of the devices to be flown, also analyzed the results of the experiments, and lobbied for continued use of the V2 rockets.

The origins and nature of this panel will be reviewed. We will concentrate upon the scientific goals of those who constituted the panel, and will identify the nature of the institutional backing each enjoyed in this activity, which may be regarded as the origin of science in space.

Karl Hufbauer, University of Calif., Irvine, & NASA Contract Historian

COOPERATIVE RESEARCH ON THE SUN, 1670-1985

Three interrelated themes dominate the social history of research on the sun since Galileo's discovery of sunspots -- (1) patronage for solar research has become increasingly more abundant, more varied in its sources, and more clearly demarcated from that for other fields; (2) participants in solar research have become progressively more numerous, more dispersed, more persevering, and more specialized; and (3) the conduct of solar research has become ever more cooperative. This paper focuses on the third trend, first tracing the rise of cooperation within solar research since the 1670s and then exploring the change in cooperationist rhetoric between its first florishing around 1900 and its revival in the space age. Although cooperation literally refers to any instance of working together, I shall restrict attention to <u>coordinated</u> undertakings involving at least two scientists or institutions.

Cooperative solar research began in 1672 with the Paris Academy's sponsorship of simultaneous observations from Paris and Cayenne of Mars' opposition for the purpose of determining the sun's parallax. Subsequent parallax expeditions were supplemented from the mid-nineteenth century by increasingly more diverse and ambitious cooperative ventures -- e.g. eclipse expeditions from the 1840s, solar observatories and coordinated observing programs from the 1870s-1880s, international organizations for solar research from the early 1900s, and orbiting solar observatories and a solar physics journal from the 1960s.

Although solar scientists cooperated at a growing rate from the 1840s, they did not really emphasize this characteristic of their work until the early 1900s. Then, with George Ellery Hale in the vanguard, they formed the International Union for Co-operation in Solar Research. For all their eagerness to promote and thereby capitalize on the cooperative trend in solar research, however, Hale and his colleagues insisted that their basic goal was to nurture "individual initiative." By contrast, some six-seven decades later, when cooperationist rhetoric reemerged, the spokesmen for solar physics no longer tempered their campaigns for patronage of large-scale projects with appeals to scientific individualism. Whatever the personal costs, cooperation had become so ubiquitous that almost all solar scientists regarded it as indispensable for sustained progress in their specialty.

Allan A. Needell

National Air and Space Museum, Smithsonian Institution

GETTING SCIENCE ON SATELLITES

On October 4, 1957 the Soviet Union successfully launched Sputnik 1, the world's first artificial satellite. America's first satellite, Explorer 1, was placed into Earth orbit on January 31, 1958. The satellites were declared part of each countries contributions to the scientific program of the International Geophysical Year.

Explorer 1 contained a device to measure ionizing radiation, two micrometeorite detectors and temperature sensors. Subsequent American satellites and deep space probes (in the Explorer, Vanguard and Pioneer series) contained ever more sophisticated instruments. Rapidly, information on the radiation and magnetic properties of near-earth and cislunar space became available, leading to important scientific discoveries -- among them the discovery of magnetically trapped radiation and details of the interactions between solar plasma and the earth's magnetic field.

Among the designers and builders of the scientific instruments intended to observe and to measure in this new physical realm, were groups based at universities, industrial concerns and government laboratories. Each group had its peculiar experience, traditions, contacts, influence, expertise, and motivations. The differences illuminate the varied contexts within which scientists responded to outside technological stimuli and to opportunities to expand their research; the results illuminate the ways in which these groups did and did not cooperate and the relative strengths of some of the forces which have shaped post-war scientific research.

The groups led by John A. Simpson at the University of Chicago and by James Van Allen at the University of Iowa came to the satellite program from quite different directions. Both had distinguished careers in cosmic ray studies. Van Allen's experience was with sounding rockets; Simpson's with aircraft and balloons. Van Allen had worked extensively with the U. S. Navy; Simpson had worked with the U. S. Air Force. Tracing the entrance of both groups into the business of "getting science on satellites" provides valuable insight into the importance of technological, political and personal factors in an emerging

Woodruff T. Sullivan, III

Astronomy Department, University of Washington, Seattle, Wash., USA

EARLY RADIO ASTRONOMY IN THE POST-WAR ERA: A COMPARISON OF DEVELOPMENTS IN ENGLAND, AUSTRALIA AND AMERICA

Extraterrestrial radio waves were discovered by Karl Jansky in 1932 at Bell Telephone Laboratories and extensively studied in the period 1939-46 by Grote Reber in Illinois. But despite this pioneering American work, it was post-war groups in England and Australia which made the essential contributions to the study of "cosmic noise" and defined a new discipline that was to transform modern astronomy. With backgrounds in physics and electrical engineering, the ex-wartime radar researchers of these groups scrambled to learn the basic precepts of astronomy even as they were permanently changing its face. British teams headed by Martin Ryle at Cambridge University, Bernard Lovell at Manchester University, and James Hey of the Army Operational Reasearch Group quickly discovered and established the nature of "radio stars", radar echoes from meteor trails, and solar radio bursts. In Sydney at the CSIRO Radiophysics Laboratory, Edward G. "Taffy" Bowen and Joseph Pawsey re-directed the work of their wartime radar laboratory into radio astronomy and likewise found themselves making one unexpected discovery after another. Yet in America, despite extensive radar development during the war and far more abundant post-war funding for science, there was extremely little activity - not until the mid-1950s were contributions comparable to those overseas forthcoming.

This paper discusses the institutional, national, technological, and scientific factors which led to the leadership in radio astronomy of Britain and Australia and to relative inactivity in the United States.

David Cahan, University of Nebraska-Lincoln

Between Academic Science and Industry: The Evolution of the Physikalisch-Technische Reichsanstalt, 1887-1914

   The opening of Germany's Physikalisch-Technische Reichsanstalt (PTR) in 1887 represents a new era in the state-support of scientific and technological research. The PTR undertook much fundamental research aimed at advancing, on the one hand, pure and applied physics, and, on the other, the metrological needs of the German state, industry, and science. In particular, this paper discusses why the PTR elected to concentrate most of its research efforts on the setting of physical standards. Two sets of needs largely determined the types of research problems undertaken by the PTR: those of a number of subdisciplines in late 19th-century physics (electricity, magnetism, optics, and heat) and those of Imperial Germany's "high-technology" industries (especially the electrical, gas, telegraphic, optical, and steel). The endeavor to meet both sets of needs directed the PTR's development into a standards and testing institute between 1887 and 1914. This evolution proved consequential for the nature of the institution in a number of ways. First, it produced a new type of physics institution, one standing in sharp contrast to academic physics institutes. Second, it produced an institution that was not fully capable of fulfilling the scientific aspirations of many of its most creative scientists.

Lillian Hoddeson* Baym

Fermi National Accelerator Laboratory, Batavia, Illinois

AGENCIES, TEAMS AND TECHNOLOGICAL DREAMS: FERMILAB'S ENERGY DOUBLER/SAVER, THE FIRST SUPERCONDUCTING HIGH-ENERGY ACCELERATOR, 1972-83

The emergence of large national research laboratories in the last half century raises for historians the important issue, among others, of the interaction between government agencies, scientists, administrators and technical problems. We explore this question in a case study of the Energy Doubler/Saver -- a pioneering high-energy accelerator and the first large-scale application of super-conductivity -- developed between 1972 and 1983 at Fermi National Accelerator Laboratory (Fermilab) in Batavia, Illinois. The Fermilab Doubler will play the historical role of the most pivotal antecedent of the 20-on-20 TeV superconducting super-collider accelerator, the "SSC," now under active design in the U.S.

*Also at the University of Illinois at Urbana-Champaign, and History Associates Incorporated.

Necah S. Furman

Corporate Historian, Sandia National Laboratories, Albuquerque, New Mexico, USA

SANDIA NATIONAL LABORATORIES: A PRODUCT OF POSTWAR READINESS

The genesis and growth of Sandia National Laboratories, the nation's largest weapons lab, stands as a pertinent case study showing the oftentimes complex, but effective interaction of government, industry, and cooperative research. Originally a part of Los Alamos Scientific Laboratory under management by the University of California—Sandia traces its roots to Z Division, a production-engineering arm located at Sandia Base on the desert outskirts of Albuquerque, New Mexico, in September 1945. Established to meet a temporary need, the organization gained permanence as an integral part of an action-reaction cycle geared to promote national readiness.

A study of this laboratory during its formative years, 1945 – 1950, shows how nuclear technology has determined policy, not only at the international and national levels, but also at the operational level. With the disintegration of US-Soviet solidarity and the collapse of attempts to establish international control of atomic energy, government leaders realized the need to replenish the nation's stockpile and refine the early atomic devices. These and other factors motivated the campaign to strengthen the nation's nuclear arsenal, which, in turn, sanctioned the continued need for the ordnance facility located at Sandia Base.

Originally conceived as a manufacturing facility, laboratory leadership soon realized that the tried and true practice of using outside contractors for production would be more viable. Hence, the pattern established by the Manhattan Engineering District was transitioned into peacetime use. At Sandia, as at other laboratories, the development of an integrated contractor complex was the result.

Despite numerous handicaps, the small laboratory showed a surprising ability to respond to national need. Within a few years, the temporary ordnance facility had been transformed into a permanent laboratory, nationally identified with weapons production, stockpiling, and surveillance—an image and association contra to that desired by the University of California management as an academic institution with a research and development orientation. When the University Regents proposed that "the best interest of government" would be served and the operation strengthened by turning it over to an industrial concern with engineering expertise, the Atomic Energy Commission and the President of the United States agreed. On November 1, 1949, the Sandia Branch of Los Alamos Laboratory became a part of the Bell Laboratories-Western Electric combine on a no-profit, no-fee basis. Thus, for Sandia National Laboratories, the postwar years—rather than representing a transformation to peacetime—were characterized by a continued mobilization of science in the name of national readiness.

Peter Lundgreen

Universität Bielefeld, Germany (FRG)

NATIONAL LABORATORIES BETWEEN GOVERNMENT AND INDUSTRY:
STANDARDIZING BUREAUS IN GERMANY AND THE U.S., 1870-1914

National Standardizing Bureaus represent the "new scientific bureaus" organizing Science in Government. A host of scientific bureaus have been founded during the late 19th and early 20th centuries. Why? My thesis is that this has been done less for promoting industry but rather for basing the regulatory state on science.

The regulatory state is defined as being in charge of matters such as public health, industrial safety, consumer protection, pollution control and the like. From these fields the case of "technical regulation" is singled out and studied in detail:

- The foundation of the German and American Standardizing Bureaus is reassessed.

- The functions of these institutions are analysed by asking, how matters of quantity, quality and safety were regulated.

- The limits of public regulation, and the modes of cooperation with private regulation are pointed out.

- The "spill-over" of technical regulation for the promotion of industry is discussed.

It is concluded that the science-based regulatory state gave birth to a functional type of organized science which is distinct from the academic science system, but also from industrial research.

Dominique PESTRE
attaché de recherches CNRS, based at CERN, Geneva.

THE BIRTH OF CERN: A TURNING POINT IN THE ATTITUDE OF EUROPEAN GOVERNMENTS TOWARDS BIG SCIENCE?

In the second half of the 40's the USA was clearly established in fundamental big science. In Europe the situation was different. While Britain had her own research programme, the countries on the Continent lagged far behind: nuclear energy was the only field investigated and the brain drain had begun.

The projects for European laboratories in Nuclear Research
These appeared at the end of 1949 and in 1950 in several places in Europe, though it was a proposal put forward by an American (I. Rabi) which served to break psychological barriers. In the following 18 months, the project was defined more clearly and a provisional organization established by the representatives of 12 European States.
A group of 30 to 40 people formed the driving force of the project between 1950 and 1952 and during the provisional period (1952-4). Its majority was composed of rather young nuclear and cosmic ray physicists, initially mainly Belgians, French and Italians, more experimentally than theoretically oriented, often in touch with big projects in their home countries. From the start a nucleus of science administrators or diplomats were part of the process, though not in their official capacities.

The attitudes of the government circles
In 1950-1, these circles showed considerable inertia, and were more than intimidated by the sums of money asked for. To avoid an immediate failure, the leaders of the projects proceeded carefully, step by step, proposing to set up first a provisional organization with a small budget. From 1952 onwards the amounts of money needed were progressively accepted by governments.
All the same, the initiative remained in the hands of the original group of 30-40 people: governments were left simply to react.

Was CERN a turning point for big science in Europe?
We would answer yes and no. No, because CERN was not the only factor which led Continental European States to fund big science: the growing awareness of this necessity had its roots in a broader set of considerations.
But yes, in the sense that the very existence of CERN accelerated this awareness: the highest levels of the State apparatus were constantly approached from 1950/51; no delay in funding CERN was admissible since it was a common venture; the acceptance of CERN led governments to free funds for national big science facilities;...

John KRIGE

University of Sussex, based at CERN, Geneva.

BIG SCIENCE AND INDUSTRY: WHO GETS CERN CONTRACTS?

This paper explores both the policy and practice whereby the European Organization for Nuclear Research (CERN) awarded contracts to industry between 1952 and about 1965. The period covers the construction of the Laboratory buildings and its two accelerators, a 600MeV Synchrocyclotron and a 25GeV Proton synchrotron. Its upper limit is set by the decision to build a new and expensive facility.

During this period orders worth almost ½billion Swiss Francs were placed. In principle they were to be restricted to firms in the dozen Member States of CERN, and were to be awarded primarily on technical and economic, not on political, grounds. Crucially, no attempt was made to apply a principle of 'just return', which would relate the money flowing back to a Member State through contracts to its industry to the relative contribution made by that State to the CERN budget. In practice this meant that firms in some States were in a particularly good position to win contracts, notably
- firms geographically close to the host state, and the host state itself, Switzerland in this case. Indeed Switzerland consistently gained 30-40% of the value of the contracts awarded during our period, though she contributed only some 3.5% of the overall budget;
- firms in countries with a strong and expanding industrial base. Here some systematic differences emerge between sectors of industry favoured - machine tools in Germany, heavy electrical equipment in Britain, and so on;
- firms in States with a strong national research programme in nuclear physics, where domestic companies had considerable prior experience in developing technologically complex items of equipment.

It is noteworthy that Europe was able to supply most of CERN's needs. Some orders were placed in the United States, in particular, but primarily for electronic equipment and computers.

Prof. Dr. phil. Karl-Heinz Manegold

Historisches Seminar Universität Hannover

DIE GÖTTINGER VEREINIGUNG ZUR FÖRDERUNG DER ANGEWANDTEN PHYSIK UND MATHEMATIK.

Die 1898 gegründete Göttinger Vereinigung zur Förderung der angewandten Physik und Mathematik bildete in Deutschland den ersten ernsthaften Versuch auf neue und folgenreiche Weise wissenschaftsorganisatorische und wissenschaftspolitische Konsequenzen zu ziehen aus der in der Phase der Hochindustrialisierung klarer erfaßten Korrelation von naturwissenschaftlich-technischer Forschungsförderung und industriewirtschaftlichem Wachstum. Mit dem Ziel, die in Deutschland institutionell getrennten Bereiche naturwissenschaftlicher Lehre und Forschung an den Universitäten mit der sich in jenem Jahrzehnt methodisch neu entwickelten technischen Forschung an den Technischen Hochschulen in engere Beziehung zueinander zu setzen, erreichte es Felix Klein, einer der bedeutendsten Mathematiker seiner Zeit, in Göttingen eine planvolle Zusammenarbeit zwischen Universität staatlicher Wissenschaftsverwaltung und industriellen Geldgebern zu realisieren und die Potentiale von "reiner" und "angewandter" Forschung personell und institutionell zusammenzuführen. Von entschiedenen wissenschaftlichen Überzeugungen ausgehend, hatte Klein, nachdem eine Reihe von Rufen an amerikanische Universitäten (Cornell, Princeton, Yale) an ihn ergangen waren, starke Anregungen von den Formen privater Wissenschaftsförderung in den USA erfahren. Dem gegenüber wurde der Wissenschaftsbetrieb in Deutschland fast ausschließlich vom Staat getragen und gestaltet, für die Wissenschaftsförderung galt allein die staatliche Kompetenz. In enger Verbindung mit dem preußischen Kultusministerium (Althoff) gelang es der Göttinger Vereinigung eine organisatorische Basis wirtschaftlicher, staatlicher und wissenschaftlicher Interessen zu entwickeln, unter der Devise "do ut des" ein wirtschaftliches und wissenschaftliches Verbundsystem von Privatinitiative und staatlichem Engagement zu erreichen, entgegen bisheriger wissenschaftlicher Tradition und staatlicher Norm. An der Universität Göttingen wurde mit Hilfe bedeutender Repräsentanten der deutschen Großindustrie (u.a. Bayer-Leverkusen, Krupp, Siemens) seit Ende des Jahrhunderts eine Reihe wichtiger naturwissenschaftlicher Institute und Lehrstühle als erste ihrer Art an einer deutschen Universität errichtet (u.a. für technische Physik, Elektrotechnik, angewandte Mathematik, Aerodynamik, Versicherungswissenschaft). Damit und durch erfolgreiche Organisation der Zusammenarbeit aller Vertreter der mathematisch-naturwissenschaftlichen Disziplinen (von den praxisbezogenen Anwendungen bis zu mathematischen Theorien - von Ludwig Prantl und Karl Runge bis David Hilbert und Hermann Minkowski) wurde die Göttinger Vereinigung im Hinblick auf den Zusammenhang von "Government, Industrie and the Growth of Cooperative Research" für Deutschland zum frühesten Präzedenzfall auf dem Weg zu den Kooperationsweisen des modernen Wissenschaftsbetriebes.

Arturo Russo

Professore Associato, University of Palermo, Italy

SCIENCE, INDUSTRY, AND GOVERNMENT IN ITALY DURING FASCISM

In the period between the two World Wars there was a significant acceleration in the process of industrialization in Italy, fostered by the social and economic conditions created in the country by the fascist regime. At the same time the government established a new institutional structure for science, stimulated by the postwar consciousness of the social and economic value of science: the reform of the university, the creation of the Consiglio Nazionale delle Ricerche, and the start of E. Fermi's research program in nuclear physics are the most important aspects of this recasting of the Italian scientific structures. The aim of this paper is to analyze the relationship between these two facets of the history of Italy in the 1920's and the 1930's, and the role of fascism in realizing and characterizing this relationship.

There are two aspects of this dual development that will be discussed. First a conception of science that distinctly separated pure from applied research, thus ignoring the role of fundamental research for industrial progress. CNR's financial support went mostly to small scale research programs in applied science, unaware of the new technological dimension of productive processes, and industry remained largely ignorant of the meaning and characteristics of modern industrial research.

The second aspect is the lack of means and facilities that lay behind the rethoric of "fascist science". Financial support for research was absolutely inadequate; no national laboratories in physics and chemistry were created; the reform of the university was based on an idealistic and humanistic conception of culture and of higher studies which failed to stimulate the formation of a diffuse scientific consciousness.

Dr. Alexander VOLODARSKY

Institute of the history of science and technology,
USSR Academy of science, Moscow, USSR

Mathematics in Ancient India

The problem of scientific influence and interaction is a major problem in the history of science.

The real contacts between Indian mathematicians and astronomers with their counterparts in Central Asia belong to the Kushan epoch (2nd c.B.C. - 2nd c. A.D.) at least. From the 3rd c. A.D. began the prolonged impact of the Indian science upon the science of Sasanian Iran. Indian scholars rendered certain influence on Syrian scholars as well. Though many astronomical and mathematical ideas of Indian scientists influenced science in Central Asia, Near and Middle East in the previous centuries, the direct penetration of Indian science began in the last quarter of the 8th century.

The Indian influence upon science in Central Asia, Near and Middle East occured in the most varied ways. Here they are: direct acquaintance with the Indian mathematical and astronomical tradition; contacts through scholars of the Sasanian Iran; contacs through Syrian scholars.

Mathematicians of Central Asia, Near and Middle East rendered substantial impact upon Indian scholars. This influence was rendered in several ways. Indian mathematicians were definitely familiar with Arabic translations and revisions of works of Ancient authors. They were also familiar with many original treatises of mathematicians of the Central Asia, Near and Middle East.

ABDI, Wazir Hasan

Project-incharge, History of Science and Technology
(NISTADS) LCUKNOW- INDIA.
Project

EUCLIDEAN GEOMETRY IN INDIA

India is amongst the earliest cradles of geometry which arose from practical needs. Over the centuries rules for computing areas, volumes of various surfaces and bodies, and also formulas for squaring the circle and circlying of the square were laid down to meet the requirement of astronomical and algebraic studies. However the main emphasis was on computation and construction.

Euclidean geometry which during the Medieval Ages had flourished in Western and Central Asia was based mainly on axioms, postulates and certain rules of deduction. Al-Berūnī attempted to introduce it into Snaskrit mathematical literature of India in the eleventh century without much success. It was only in the 17th century, that Jagannāth Samrāṭ, and associate of Jaisimha rendered Naṣīr-al-Dīn Ṭūsī's Taḥrīr-i-Uqlīdis into Sanskrit as Rekhā Gaṇita but the translation does not appear to have attracted much attention.

On the other hand Taḥrīr-i-Uqlīdis had become the main source of geometrical knowledge for mathematicians whose medium was Arabic or Persian. Some books were included in the Syllabi at centres of advanced learning. Accordingly many commentaries (especially on Books I-IX) were composed by various writers like Barkat b. ʿAbd-Al-Raḥmān, Ḥasan b. Dildār ʿAli, Khair Allāh Khān, the earliest being Moḥammad Hāshim (16th. century). However Book X did not attract much attention of these commentators which deals with irrational numbers.

S. M. Razaullah Ansari

Professor, Department of History of Medicine & Science, IHMMR, New Delhi

ASTRONOMY IN MEDIEVAL INDIA

Two traditions are clearly discernable in the development of astronomy in India. The ancient Indian (Hindu) calendaric astronomy as extant in the text of <u>Vedāṅga Jyotiṣa</u> was developed into planetary astronomy in the first few centuries of our era, when astronomical treatises in Sanskrit (the <u>Siddhāntas</u>) were written. In Medieval India (15 - 18th centuries A.D.) mathematics and astronomy as developed in the Islamic Civilization of West-Central Asia, particularly that of the later period /i.e., from <u>Nasīruddīn Al-Ṭūsī</u> (1201-74) to <u>Bahāuddīn Al-ʿĀmilī</u>, (d. 1622)/ with its emphasis on the organised observational/practical astronomy in the sense of the establishment of observatories and compilation of <u>zijes</u> (the tables), was brought to the Indian courts of Mughal Emperors by Central-Asian Muslim scholars. It is therefore natural that Mughal rulers promoted astronomy in Medieval India, not only that of their own tradition but also interacted with the Ancient Indian (Hindu) tradition, though this fact is not widely known.

A few characteristic features of the Indian Medieval period are:
1. Translation of astronomical/astrological texts from <u>Sanskrit</u> into Persian, and vice versa particularly in the later Mughal period,
2. Efforts to establish astronomical observatories in the reigns of <u>Humāyūn</u> (1530-36), <u>Shāhjahān</u> (1627-1658), which culminated in the actual foundation of a number of observatories in various Indian towns in the time of <u>Muḥammad Shāh</u> (1719-1748), and 3. Compilation of about a dozen original zijes.

In this paper we discuss fairly in details these features of Medieval India astronomical activity. We further substantiate that the first Indian zij (<u>Zij-i Nāṣirī</u>) was compiled about 1265, even before <u>Al-Ṭūsī's</u> <u>Zij-i Ilkhānī</u> (ca 1272) and the first observatory was supposed to be built in 1408, even before <u>Ulugh Beg's</u> observatory at Samarqand (ca 1420).

SUBBARAYAPPA, B.V.

Centre for History and Philosophy of Science, Indian
Institute of World Culture, Basavangudi,
Bangalore 560 004 (INDIA)

TECHNOLOGY IN INDIA UP TO 1750 A.D : A PERSPECTIVE

The Indus Valley civilization (fl 2350 BC-1700 BC) was noted for a wide variety of techniques such as glazed ceramics, terracotta, standardized brick production, spinning and weaving, bead-making, metal-working specially of copper-bronze and cire perdue process.

Iron made its appearance quite some time after this civilization atrophied, and glass still later. Gold, silver, copper and iron had their social dimensions, the last becoming an important component of the Megalithic culture (c 600 BC - 200 AD). The technology of iron and copper assumed macro-levels - pillars and icons - from about the 5th cent. A.D in tune with the social and religions imperatives. The jewellery craft, textiles, cosmetics and perfumery elevated the quality of living. The craftsmen were able to become part of the guilds and thus participate in commercial enterprises.

Admittedly, the craftsmanship knew no inhibition, nor geographical barriers. Indian craftsmanship, specially in the medieval period (c 1200 - 1800 AD) developed several growth-points like paper-making, carpet weaving, intricate metal working, pyrotechnics, gun powder and rocketry, marble and jade carving, through assimilation of techniques from outside. The over-land as well as sea trade routes and India's geographical position vis-a-vis these routes, played an important role in the assimilation and transmission of several techniques.

Certain technical professions were prone to result in a type of stratification of castes or sub-castes, in turn, giving rise to ingrained elements of regressive character. The paper attempts to present a general profile of technology in India in a historical and social perspective.

Xi Zezong

Professor and Director of Institute for the History of Natural Science,
Academia Sinica, Beijing, China

A study of Tunhuang calendar

   This paper collects thirty-two items of Tunhuang manuscripts of
calendar from the 9th to the 10th century which are now preserved
in the Beijing Library, the British Museum and the Bibliotheque
National in Paris. Only a few of them clearly indicate their dates,
such as 877 A.D., 982 A.D., etc. The most are fragments. Here we
try to date these fragments based on the astronomical as well as
the astrological contents in them and to compare them with the
calendar used in Central Plains at the same time.

Michio YANO

International Institute for Linguistic Sciences, Kyoto Sangyō Univer[sity]

Professor of Sanskrit and Linguistics

WESTERN ASTROLOGY IN T'ANG DYNASTY CHINA AND ITS SURVIVAL IN JAPAN

Numerous texts on astronomy and astrology of Western origin were brought to China in T'ang Dynasty by such variety of religious groups as Buddhists, Zoroastrians, Nestrians, Manicaeans, and Muslims. Although many were translated into Chinese, only a small number of them survived. What few texts are available to us belong to Tantric Buddhism, except the Chiu-chih li (九執曆) which represents the Ārdharātrika school of Indian astronomy fairly well. Some of the texts were accessible to Japanese Buddhists who visited China in T'ang dynasty. The most famous one was the Hsiu-yao ching (宿曜経), translated into Chinese by Pu-k'ung in 759, which was brought to Japan by Kūkai (空海) in 806 and which served as the basic text of the new school of astrology called Sukuyō-dō. Another interesting text, the Ch'i-yao jang-tsai-chüeh (七曜攘災決) belonging to the end of the 8th century, was introduced in Japan in 865 by a Buddhist monk Shūei. This text provided planetary ephemerides. These phemerides, despite their crudeness, were in use for the purpose of casting horoscopes. The text, however, soon gave way to another peculiar text, Fu-t'ien li probably of Central Asian origin, whose small fragment survives only in Japan. In the list of books brought from China by Shūei, we find the title Tu-li-yü-ssu ching (都利聿斯経) which, although no longer existing, seems to have been an important reference book for Japanese astrologers in their interpretation of horoscopes. After checking the few lines which survive as quotations in Japanese horoscopic documents, I am inclined to think that the title is nothing but a transcription of the name Ptolemaios, just as the other title Ssu-men ching (meaning a book consisting of four chapters) is a translation of 'Tetrabiblos'. It is interesting to observe that astrology of Western origin, through the modification of Tantric Buddhism, had a better chance of survival in Japan than in China.

PARK, Seong-Rae

Professor, Hankuk University of Foreign Studies

TRANSMISSION OF SCIENTIFIC IDEAS AND LANGUAGE

   Modern scientific ideas were transmitted into East Asia via translations. Korean experiences, in comparison with the Japanese and the Chinese, are particularly interesting and illustrating.
   The Chinese and the Japanese started to have contacts with the West from the early seventeenth century through the visiting missionaries like Matteo Ricci and Francisco Xavier among others. Yet Koreans had never had any direct contact with the Westerners on intellectual level until the nineteenth century, for Western missionaries did not come. Under the given circumstances the only way of learning science before the nineteenth century was through the imported books from China. Western missionaries translated many books on science into Chinese there.
   Koreans started their study of science through the Chinese rendition of Western science, not through Western books. The tradition lingered on even after the opening of Korea in 1876. Though Westerners were formally accepted into their land, Koreans never bothered to learn any of the Western languages as a means of absorbing science. Instead they had eventually switched their transmitter from the Chinese to the Japanese at the turn of the century. The basic pattern of indirect transmission of scientific ideas continued in the Japanese language period, 1900-1945.
   Direct transmission of scientific ideas into Korea became finally possible only after 1945. Very soon, however, another difficulty developed for free transmission of scientific ideas among the three peoples of East Asia. Activated by the post-war nationalism, the Chinese started to build their own terminology in science and the Koreans simply stopped their use of Chinese characters.
   What we see today is the gradual destruction of Chinese character-based communication system in East Asia as an intermediary language. It is time that East Asians have to start to reestablish something for the losing vitality of their language system for more effective transmission of scientific ideas among the three peoples. Efforts for the device of common scientific terms are highly desirable.

Shigeru NAKAYAMA
University of Tokyo

Japanese response toward incoming Western mathematics

Located at the eastern end of the Eurasian world, Japan received Western impact successively in the following five waves: 1) direct contact with Western navigators up to the early seventeenth century, 2) Influence of the Sino-Jesuits writings in the eighteenth century, 3) the translation of imported Dutch texts since the late eighteenth century, 4) direct schooling by Dutch navy officers at Nagasaki in 1850s and 5) the Chinese translations of Protestant missionaries' writings after the Opium War.
Mainly because of the pre-existence of traditional Japanese mathematics, *wasan*, the influence of Western mathematics was less significant compared to those of medicine and astronomy. As the heart of Japanese mathematics was playful and artistic divertissement, the characteristics of Western mathematics was complementarily considered to be practical to serve for navigation and surveying. The meaning of Euclidian geometry was totally missed, as the Japanese thought it too primitive. The clearest demarcation between traditional and Western mathematics was the Japanese use of the abacus while the Western one depended on calculation by paper and pencil. Secondly, those who practice of Western mathematics were the people who wrote algorithm in terms of Western numerals and horizontal arrangement of formulas while traditional mathematicians adhered to the vertical expression of mathematical formulae.
It is often claimed that the Japanese could receive Western mathematics easily because of their preexisting background in Japanese traditional mathematics.
I am not denying this argument but take it as a meaningless statement. Both school, traditional and Western, had their own paradigms so that it was extremely difficult to switch from one school to another.
Thus, paradigm shift from traditional to Western was completed only when the old traditional generation was replaced by newly educated mathematicians.

Ubiratan D'Ambrosio

Professor of Mathematics, Universidade Estadual de Campinas

The problematics of History of Science in Latin America

We assume a broader than usual conceptualization of Science which allows for looking into common practices which are apparently unstructured forms of knowledge. This comes from a concept of culture which is the result of an hierarquization of behavior, from individual through social behavior and leading to cultural behavior. This dependes on a model of individual behavior based on the ceaseless cycle ...reality→individual⇄action→reality.... The conceptualization of Science which derives from this model allows for the inclusion of what might be considered marginal practices of a scientific nature, and which we may call ethnoscience.

The basic question may be posed as follows: how theoretical can ethnoscience be? There are numerous practices among different cultural groups in Latin America, going on through generations, which fit into this broad concept of ethnoscience. This has been recognized frequently by anthropologists and even before etnography has been recognized as a science by several travellers. We go a step further in trying to find an underlying structure of inquiry in these ad-hoc practices by posing the following questions: 1. How to pass from ad-hoc practices and solution of problems to methods? 2. How to pass from methods to theories? 3. How to proceed from theories to invention? By looking into the History of established Science, these seem to have been the crucial steps. The main issue is then a methodological one. Current approaches to the History of Science tend to answer the three basic questions above within a framework which results from the structured form of knowledge which is at the same time the object of analysis. Historical analysis becomes then a merely descriptive review of theory itself. Consequently, this excludes knowledge which is not enbodied in the theory itself, such as ethnoscience, depriving it of any history, hence of the status of knowledge. These are the issues we propose to face.

Hebe M.C. Vessuri

Head, Science and Technology Department, CENDES-UCV, Caracas

The modern implantation and development of scientific disciplines in Venezuela and their social implications.

Scientific disciplines constitute structuring units of intellectual activity, entities that incorporate in an organic fashion intellectual, institutional and personal elements, in order to give form to the reality perceived and experimented by individual sceintists.

The paper maintains that the sociohistoric study of scientific disciplines advanced in the central countries is enriched when the specificity of the phenomenon of its implantation and development in peripheral societies is considered. In this latter context, the conditions of the transfer of epistemological models and disciplinary contents and the means used in the transfer, as well as in the reception, interpretation and utilization of those knowledges in the receiving society, taking into account the interactive nature of knowledge exchanges and therefore the local participation in the reproduction and development of particular disciplines and specialties, are new elements brought to bear in the analysis of scientific disciplines, allowing to ask new questions and, hopefully, helping it to acquire a special meaning.

More specifically, the paper aims at presenting some of the results of the volume just published inSpanish on "Academic science in modern Venezuela. Recent history and perspectives of scientific disciplines" under thr author's editorship, by the Venezuelan Association for the Advancement of Science (Aso VAC). On the basis of fourteen papers that make up the volume, which were specially commissioned and which involved extensive research on specific disciplines, the social and cultural dimensions of the process of transfer of knowledge inthis developing society are explored.

Antonio Lafuente

L'EXPEDITION 'LA CONDAMINE' (1736-1743) ET LES ACTIVITEES SCIENTIFIQUES ET TECHNIQUES DANS LE VICE-ROYAUTE DU PEROU

En 1736 il arrive à Quito l'expédition géodésique hispano-française organisée par l'Académie royale des Sciences (Paris) pour la détermination de la valeur d'un degré de meridien à côté l'Equateur. Pendant quelques années cette ambassade singulier va vivre ensemble avec des institutions et des hommes de science du vice-royauté du Perou. L'objet de cette communication sera monstrer l'incidence de l'expédition dans le procés de relancement et de légitimation du travail qu'on etait en train de faire quelques créoles éclairés, et aussi dans le développement des activitées scientifiques dans les domaines de la médicine, la botanique, la géographie y la génie civil et militaire. Nous nous proposons également d'étudier leurs répercussions en Espagne et les relations que les expéditionnaires ont eu avec l'Administration coloniale espagnole. On peut dire que tant dans la métropole que dans l'Audiencia de Quito et Lima, ce contact avec la science française fut un jalon significatif dans le procés d'assimilation de la science moderne.

Edmundo F. Fuenzalida

Associate Professor - Stanford University

Scientific Research in Chile: from the 19th to the 20th Century

    The history of scientific research in Chile since Independence has still to be written. There are contributions to it in the different general histories of the country, such as those by Barros, Arana, Encina, Heisse and Vial. There are also monographs on individual researchers and specific scientific institutions, such as the Museum of Natural History. In spite of the scarcity of particular studies, it is possible to characterize the type of scientific research that was undertaken in the country during the 19th century. The transition to the 20th century happened when this century was already well advanced. But this delay is compensated by a concentrated effort to create a National Scientific establishment. The effort is led by the National University, but the other universities and the National Government join into it very soon.

    Scientific Research in Chile in the 20th century shows characteristics that distinguish itself clearly from that performed in the previous century. In spite of the efforts of leaders, government officials and scientists, what had begun as an effort to lay the foundation of a <u>national</u> scientific establishment, produced the outcome of a segment of todays <u>transnational</u> institution of scientific research and development.

LUIS CARLOS ARBOLEDA A.

Profesor, Depto. de Matemáticas, Universidad del Valle, Cali, Colombia

RECEPTION DES FONDEMENTS DE L'ANALYSE EN COLOMBIE

L'étude de l'évolution de l'enseignement du calcul en Colombie, présente deux moments d'intérêt historique. D'une part, l'introduction dans les années 1760 du calcul de Newton y Leibniz dans la première chaire de mathématiques professée par Mutis, Directeur de **l'Expedición Botánica del Nuevo Reino de Granada**. Les textes les plus représentatifs de sa conception plutôt éclectique furent ceux de Wolff et Bails. Puis, les années suivant la création en 1847 du **Colegio Militar** où circulèrent dans le pays quelques uns des traités de la période de la préfiguration de la rigueur à la Cauchy: ceux de Lagrange, Laplace, Legendre et, surtout, celui de Lacroix. Sous l'influence de ces traités ont été rédigées les premières publications mathématiques faites en Colombie par Bergeron, Pombo, Liévano el leurs éleves.- Un enseignement de l'analyse exprimant une volonté de exposer en forme claire et précise les notions fondamentales, seulement comence à se proffesser en Colombie à la fin du XIXe siecle. Malheureusement l'expérience fut avortée avec les événements de la dite **Guerra de los mil días** (1898-1902), la dernière d'une série intermittente de neuf guerres civiles, deux guerres internationales et plusieurs dizaines de révoltes régionales qui avaient marqué la vie politique et intellectuelle de la jeune république depuis 1810. Coïncidant avec une période de stabilité sociale et de renouvellement de l'Université par l'Etat, autour des années 1880, petit à petit s'impose chez nous la tradition d'enseigner le calcul infinitésimal dans nos écoles d'ingénieurs. Néanmoins, parce qu'il tient ses origines dans des institutions plutôt techniciennes, cet enseignement va se caractériser par son orientation pragmatique. L'instruction devait porter sur des savoirs mathématiques censés être d'utilité inmédiate. Une formation théorique plus avancée détournait l'attention des ingénieurs de la solution des besoins pressants du développement auxquel ils devaient se consacrer. En tout cas, on sait que déjà en 1880 circulent dans le pays des textes d'analyse telles que ceux de Sturm, Serret et Laurent. La lecture de ces textes assurément a renouvellé et approfondi, au moins dans quelques uns de nos **ingenieros-matemáticos**, l'intérêt pour les questions des fondements, en les poussant même à mener des études précises. Toutefois, son influence est restée limitée à l'initiative personnelle. La situation changera de fond en comble, avec les réformes de la période d'après-guerre et la création d'institutions mathématiques modernes et professionnelles.

Rafael Chabrán

Assistant Professor, Louisiana State University, Baton Rouge,

THE RECEPTION OF DARWINISM IN ARGENTINA: AN APPROACH

The reception of Darwin's ideas in Argentina has received very little attention. Working within the guidelines provided by Thomas F. Glick ("Perspectivas sobre el Darwinismo en el mundo hispano," 1984), I outline an approach to the study of the reception of Darwinism in Argentina.

The introduction of Darwinism in Argentina was tied to the coming of Spencerian positivism. The Generation of 1880 embraced Spencer's agnostic and evolutionary positivism. The group played a dominant role in Argentinian educational life.

Argentina had an established scientific tradition in paleontological research which contributed to the reception of evolutionary thought. Most significant was the work of Francisco Javier Muñiz(1785-1871). He established the first important paleontological collection in Argentina and founded the Associación de Amigos de la Historia Natural.

One of the principal centers for the diffusion of evolutionary thought was the Escuela Normal de Paraná (1870). Pedro Scalabrini(1848-1916) was influential in spreading evolutionary doctrine from this institution. He sought to replace colonial scholasticism with the scientific spirit of evolutionary positivism.

Florentino Ameghino(1854-1911) was one of the foremost defenders of Darwin in Argentina. Ameghino studied Lyell and Darwin's works in the 1870s. In one of his most important works, Filogenia (1884), he sought to prove evolutionary thought, as well as to put forth a system of evolutionary classification based on natural laws and mathematical proportions. Throughout his life, Ameghino engaged in polemics against anti-Darwinians, especially the creationist, Hermann Burmeister(1807-1892).

E.L. Holmberg(1852-1937)was another important popularizer of Darwin's ideas. While trained in medicine, Holmberg dedicated his life to natural history and literature. One of Holmberg's clearest defenses of Darwin can be found in his short novel, Dos partidos en lucha (1875).

The impact of Darwinism on the development of Argentine social sciences can be studied by considering the work of José Ingenieros(1875-1925), one of the acknowledged leaders of positivism and scientific thought in Argentina. Ingenieros, deeply influeneced by Ameghino, used evolutionary models in his writings.

Rosaura Ruiz Gutierrez

Professor   Facultad de Ciencias UNAM

Introduction of Darwinism in Mexico
1) The haeckelism in the mexican biology

Ernst Haeckel is known as the greatest diffuser of darwinism. One of the books about evolutionism more expended on its numerous translations is his History of the creation - (1868). Haeckel presents one version exceedingly simple - and moreover adulterate of the darwinian theory of evolution. His importance lies on two matters, because the haeckelian interpretation is incorporated in so many countries.

This paper attempts to explain the case of Mexico, where Haeckel influence is remarkable, since it concerns not only the knowledge of the evolucionist theory but also has relation with the biology constitution.

At the end of the XIXth century other type of research begins being its objective the study on the most general problems that in fact distinguished the biology, be example those of Alfonso L. Herrera about the origin of life. These works and some others are characterized by their antivitalism. Their conception of unity from all the Universe phenomenon, and they identify between the inert and living beings agree with the german Naturphilosophen, transformed by evolutionism of Haeckel.

Eduardo L. Ortiz

Imperial College, London, England

EINSTEIN'S VISIT TO ARGENTINA IN 1925

In the first quarter of this century the exact science received considerable attention in Argentina. A large Physics and Astronomy complex was built in the city of La Plata in the 1910's and similarly important changes were begining to take place in Mathematics at Buenos Aires University.

From the early 1920's the possibility of inviting Albert Einstein to lecture in Argentina was discussed in Bunos Aires. Several factors contributed to give momentum to these initiatives. The period was one of intense intellectual debate in Argentina. Suffrage was granted to a larger proportion of the population and Universities were experiencing fundamental changes through the so-called "University Reform" movement.

Interest on Relativity Theory had already crystalized in a few books where the theory was discussed at different levels. However, even in the most mathematically minded, the philosophical element is clearly perceptible.

The figure of Einstein was associated with pacifism and with ideas of revolutionary change. Einstein's visit to Spain helped to make this perception even more precise.

A large and succesful Jewish community was then trying to find a place in Argentina' society and joined forces with the promoters of Einstein's visit. The equally large and important German community in Argentina was divided on this issue, as on several others, at the time.

The visit marked the close of a period of important change in the intellectual life of Argentina. Is background, development and aftermath is discussed in this paper.

Nancy Leys Stepan

Associate Professor of History, Columbia University

LATIN AMERICAN SCIENCE AND THE WORLD EUGENICS MOVEMENT

This paper argues that the emerging new wisdom about the history of genetics and eugenics is inadequate because of the systematic neglect of genetics and eugenics in the Latin American context. Looking at eugenics as a world movement, rather than one defined by the historical experiences of selected European and North American cases, eugenics in Latin America emerges as a major variant in its own right, one that will help redefine eugenics as an international, scientific movement of great variability and complexity.

The new wisdom about genetics and eugenics is based on close study of the United States, Britain and selected European cases. This work emphasizes that scientifically, eugenics was closely associated with the rise of Mendelian genetics. Socially and structurally, eugenics was associated with pessimism about the failures of social legislation to cure social ills. From a policy point of view, because Mendelian eugenicists rejected Lamarckian ideas of heredity and drew a sharp distinction between "nature" and "nurture", the eugenicists emphasized the importance of selection and differential breeding of the "fit" and the "unfit" in the improvement of the human species. Eventually, birth control, segregation and sterilization of the "unfit" became the social tools of the eugenicists.

Eugenics, in short, is seen as having a predictable scientific style, and though it drew into its ranks socialists, liberals and conservatives, its policy implications were also predictably "anti-environmentalistic."

It is this scientific and social style of eugenics and genetics that is challenged by a consideration of eugenics and genetics in Latin America between 1900 and 1940. Particular attention will be paid to eugenics and genetics in Brazil, Mexico and Argentina. Scientifically, in its origins many of the eugenics movements in Latin America derived from French models and tended to be initially Lamarckian rather than Mendelian in genetic style. It was Lamarckian conceptions of inheritance that structured the debates in Latin America about the effects of environmental or racial "poisons," such as alcohol, on the hereditary constitution of human populations. Socially, eugenics was not a response to the supposed failure of social and sanitation legislation to improve the physical, mental and moral conditions of the poor. On the contrary eugenics was a call for the introduction of "modern" social legislation and influenced the form that legislation took. In particular, it drew attention to the supposed effects of disease in reproduction and led to calls for "puericulture" (child care before birth) and to "prenuptial" examinations as state-sanctioned impediments to marriage. These elements of eugenics and eugenics came together in the Latin Eugenics Federation, founded in Mexico in 1935 with delegates from fifteen Latin American countries and from France and Rumania. The existence of Lamarckian, racialist eugenics as well as Mendelian challenges to racialism within the Latin American eugenics movement adds further interest to the Latin American cases, indicating that the inherent logic of science alone does not determine its social meanings and outcomes.

SABRA, Abdelhamid I.

Professor of the History of Arabic Science, Harvard University

The Intra-cultural Perspective in the Historical Study of Arabic Science

The purpose of this paper is to make a plea for adding a new dimension to the historical study of Arabic science: namely, a serious consideration of the relevant factors of Islamic civilization. This is not an original plea (Aydin Sayili's The Observatory in Islam, 1960, is one major study which implements it), but it has so far been generally ignored, avoided, or, at times, even slighted, except when the seductive question "Why did Arabic science decline?" is being discussed. The wholesale transfer of Greek science to Islam in the eighth and nineth centuries was a most important event in the history of culture, with far-reaching consequences for the history of Greek learning, for Islamic civilization, and for the later history of science in Europe. But this event has been too often viewed as an intermediate stage in a journey which occurred in a rather neutral space. While part of the story can certainly be told in kinematical terms (the bare movement and miraculous transformation of scientific products--texts, theorems, techniques--to which are attached spatio-temporal symbols in the form of personal names, dates, and geographical locations), the outcome cannot by itself make a satisfactory story. Greek science did not automatically turn itself into an Arabic enterprise, nor did the scholars who appropriated it aim to satisfy the needs of anyone but themselves and their contemporaries. Thus no understanding of the events of those two fateful centuries is possible without taking into account the forces, ideals and concerns of the burgeoning Islamic civilization. Equally obvious remarks apply to an understanding of the development and character of the scientific tradition subsequently maintained in Medieval Islam. This is not a plea for essentialist explanations, and no elixirs are being prescribed. The one simple fact being premised here is that Arabic science _was_ a phenomenon of Islamic civilization. It is not extravagant to recommend that it should be studied as such.

F. Jamil Ragep

Postdoctoral Fellow, Harvard University

Theories of Trepidation from Antiquity to Copernicus

   The theory of trepidation, despite (or perhaps because) of being both strange and erroneous, can reveal a great deal about the nature of the transmission of scientific ideas from one culture to another. From its probable origination in the second century B.C. until it found a place in Copernicus's De revolutionibus, the theory underwent considerable changes as it passed from Greece to Islam to the Latin West. These changes are indicative of the attitudes of astronomers both within and between different cultures toward the problems of the acceptability of astronomical models, the testing of these models, and the reliability of received observations.
   From the report of Theon of Alexandria (4th c. A.D.), one learns that the original theory called for the solstitial points to move in one direction for 640 years and then reverse direction for the next 640 years. This theory, perhaps proposed as an embellishment or alternative to Hipparchus's discovery of precession, was ignored by Ptolemy who adopted a simple monotonic motion. Given the great prestige of the Almagest as well as the fact that the solstitial points did not change direction in 483 A.D. as trepidation required, one would have assumed that by the Islamic period the theory would have expired unmourned. But the faster rate for precession found by Islamic observers as well as a different value for the obliquity of the ecliptic from that given in the Almagest led to a variety of proposals that would account for either the first or both of these changes. The ancient theory of trepidation, along with its parameters, was resurrected to become part of these new theories. The best known of these, and one that had a profound influence in the Latin West, has long been mistakenly attributed to Thābit ibn Qurra (d. 901 A.D.). In fact it is his grandson Ibrāhīm ibn Sinān who is most closely associated with trepidation in the eastern Islamic tradition. The two models by pseudo-Thābit and Ibrāhīm offer contrasting approaches to celestial motion; the former depends on techniques of Ptolemy's latitude theory while the latter uses the more physically acceptable methods of Eudoxus.
   By the later Middle Ages Islamic astronomers continued to discuss trepidation as an interesting mathematical problem, but most had come to regard it as implausible. With the passage of centuries, Ptolemy's values for precession and the obliquity began to appear more and more suspect. Interestingly enough, the theory of trepidation in the Latin West was taken rather more seriously, perhaps because Ptolemy's parameters were not as readily challenged.

Khalil JAOUICHE

Chargé de recherche au C.N.R.S. (Paris)

ANALYSIS AND SYNTHESIS IN ISLAMIC MATHEMATICS: THE BOOK OF IBN AL-HAYTHAM.

The method of Analysis and Synthesis has played a very important role in Ancient and Medieval mathematics and many examples of its application can be found in ancient greek and arabic books. But there have existed very few texts dealing with this subject from a general point of view, either in greek or in arabic. All the modern researches on this problem rely almost exclusively on the short introduction of Pappus' Book VII of the <u>Collection of Mathematics</u>. But the most important books - because the most lengthy and the most exhaustive - which deal with this problem are Ibrahim ibn Sinan's (609-946): <u>The Methods of Analysis and Synthesis</u>, and Ibn al-Haytham's (965-1041): <u>Book on Analysis and Synthesis</u>, the edition of which we have now achieved together with a french translation. The first one is pedagogical and intends to teach students of high level how to practice analysis and synthesis. The second is more theoretical and gives a very complete study of the subject. In my paper I shall give, particularly, a brief study of the most important features of this latter book:
1) The fields which it covers (arithmetic, geometry, astronomy, music); 2) Classification of the problems studied by the method of analysis and synthesis: arithmetical, geometrical, determinate, undeterminate, limited by a condition, not limited by a condition; 3) The problem of auxiliary properties and constructions and the relation of these with intuition together with their role in the discovery of demonstrations and constructions in theorems and problems; 4) The problem of the direction of logical inference.
This general and brief study will of course be sustained by summarized examples drawn from the book of Ibn al-Haytham.

E. S. Kennedy

Institut für Geschichte der Arabisch-Islamischen Wissenschaften, Frankfurt 1, B. R. D. (Professor Emeritus, American University of Beirut)

MEDIEVAL CHINESE AND UIGHUR WORDS IN IRANIAN ASTRONOMY

From ten manuscripts which comprise one or more copies of seven Persian astronomical handbooks (zījes), a collection of 139 non-Islamic technical terms has been compiled. Of these, thirty-two are in Uighur Turkish, the rest are Chinese. The late Professor Joseph Fletcher undertook to infer the words in the original languages from the transcriptions in the Arabic characters. The task is complicated by the large number of variants in the sources inadvertently introduced by successive scribes ignorant both of Turkish and Chinese. However, for the elements of the standard calendrical cycles the solution turned out to be straightforward.

These include the duodecimal (animal) cycle for double hours, and the twelve month names, in both Chinese and Uighur. The decimal and duodecimal components of the sexagesimal cycle and the names of the twenty-four half months are in Chinese only.

The remaining words name concepts used in the distinctively Chinese practice of approximating solar and lunar true longitudes by (in our terms) arcs of parabolas. Several of these words remain unidentified.

The earliest of the sources is the Zīj-i Ilkhānī (c. 1270) by Naṣīr al-Dīn al-Ṭūsī, the director of the Mongol observatory at Marāgha in west-central Iran. At least one Chinese scholar is known to have been associated with Naṣīr al-Dīn, a certain Fu Mêng-chi (?). Presumably he instructed his Muslim colleagues, whence the extensive tables and intricate procedures were faithfully transmitted in Persian for at least two centuries.

Stephen C. McCluskey, West Virginia University, Morgantown, WV USA

CALENDARS AND SYMBOLISM: FUNCTIONS OF OBSERVATION IN HOPI ASTRONOMY

The rituals of the Pueblo Indians of the Southwestern United States incorporate a number of astronomical concepts and motifs. In particular, their ritual calendar is astronomical in nature and is maintained by careful observations of the motions of the sun along the horizon and, to a lesser extent, of the phases of the moon. The solar observations also define a sacred coordinate system based on the four solstitial directions, which appear frequently as a part or ritual and also provide a reference frame for the observations themselves. The differences among sites used for astronomical purposes by the Hopi Indians of Northern Arizona indicate that there are two types of ritual observations that serve different purposes and employ different kinds of markers.

Calendric or anticipatory observations are made prior to major rituals, especially prior to the solstices, by an elder of the religious society sponsoring the coming ritual. The observing techniques are generally highly precise, typically employing extremely long sightlines (50 to 100 km) extending to natural landmarks on the distant horizon. The exact interval from the observation to the ritual is not fixed, but is determined in part by the existence of a suitable landmark where the Sun rises or sets prior to the desired ritual.

Symoblic or confirmatory observations are made during rituals associated with the winter and summer solstices when, as is widely known, the Sun pauses at his house. The sightlines are generally short (about 10km) and lead to shrines erected on the horizon that are too small to be seen from the observation point. These observations cannot be very precise since the distance is short, the marker is not visible, and at the solstices the sun is moving slowly. These shrines mark where the sun rises or sets at the solstices and serve more as sites where rituals are performed and offerings are deposited for the Sun than as precise observation markers.

These two classes of sites appropriately reflect the different functions of these two classes of observation. Anticipatory observations provide the elders with data needed to perform their scientific function of maintaining the calendar. Confirmatory observations empirically confirm for society at large, including the elders, not the accuracy of the preceding observations but more fundamentally the symbolic framework of four sacred directions which provides the theoretical basis for Pueblo astronomy.

Johanna Broda, Universidad Nacional Autónoma de México, Mexico

GEOGRAPHY, CLIMATE AND THE OBSERVATION OF NATURE IN PREHISPANIC MESOAMERICA

Understanding and uses of nature in Prehispanic Mexico are discussed in terms of the exact observation of the natural environment and the geographical and climatological notions about the "Known World". The latter concept implies the dimension of socially delimited space. Material from the Aztecs of Central Mexico is presented, but comparative archaeological, ethnohistorical as well as modern ethnographic examples from other regions of Mesoamerica are also given. The data refer to the cult of mountains, caves, sacred wells and springs as well as the sea. They also include references to agricultural cycles, the notion of the seasons, calendrics and the body of astronomical knowledge.

The contribution of this paper comes from the field of Anthropology although, at the same time, an interdisciplinary and comparative approach is considered necessary. The focus of interest deals with the dialectical relationship which Ancient Mesoamerican cultures established between the precise observation of nature, a body of scientific knowledge and its shading off, at a certain point, into cosmology and ritual.

Aztec cosmovision was a system which coherently explained the known universe in terms of precise knowledge as well as the ideological needs of that society. In general terms, it is inferred that the conception which a society forms itself of nature is a reelaboration in social consciousness of the conditions of the natural environment. The latter are never perceived in exactly the same form in different societies; there does not exist a "pure" perception separate from the social institutions which produced it.

A further point of the investigation refers to the striking survival which certain elements of this cosmovision have shown up to the present day. While the cult of water, mountains and the earth formed part of the ideological expression of Prehispanic State Cult, after the Spanish Conquest it lost this integration into the official cult. However, these elements survived vigorously due to their vinculation to the necessities of peasant life. From an expression of elite culture they transformed themselves into the cult of Indian peasant communities thus losing their articulation with the wider coherent ideology of a previously autonomous society. Keeping in mind these structural changes over time, the modern ethnographic data permit to gain extremely valuable insights which complement our data for reconstructing Prehispanic cultural concepts.

Franz Tichy

Prof. Dr. phil., Universität Erlangen-Nürnberg, Erlangen
Fed. Rep. of Germany

THE HORIZON REFERENCE-SYSTEM. ITS PERCEPTION AND APPLICATION IN THE
ORDER OF TIME AND SPACE IN MESOAMERICA

Mesoamerican man living in the center of his cosmos observed the movement of the sun and other celestial bodies. As reference points he recognized the extreme positions of the sun on the horizon at the time of the solstices giving him his symbol of the sun. At latitude 19°N the range of horizon covered annually by sunrise and sunset points extends to 49.5°. East and West as well as North and South he was not able to define as precisely. The sun is near the zenith at noon, the North celestial is too close to the horizon to be recognized. On their long meridional migrations the Indian peoples probably observed the different paths of the celestial bodies in the different latitudes, and in the tropics the cenital positions of the sun and the pleiades. The caves of Xochicalco and Teotihuacan surely were utilized as observatories.

The observation made from a particular site, had to be detached from topographical relations such as their relative position to mountains. Thereafter they could be transferred through noting the situation in the horizon reference-system or with the day in the calendar. However, we do not know any instrument to measure the arcs on the horizon or on the meridian in Precolumbian times. Probably the thumb jump or a tendril had to be sufficent. The often identical orientations of the axes of pyramids and churches in Mexico led to the conclusion that there exists a general unit of 4.5°, i.e. 1/20 of the right angle. There are 11 units between the solstitial points on the horizon. The extreme distance of the sun at noon at the summersolstice with 4.5° to the North at Cholula or at Xochicalco (Lat. 19°) or 9° at Guatemala (Lat. 14.5°) might have been observed. Could this have been the beginning of science?

Mesoamerica was in posession of the sun calendar with 365 days but without leap days. Nevertheless there may also have existed fixed calendars for the application in agriculture and agroreligious purposes. There is the agrarian cycle of 260 days, fixed by means of the cenital positions of the sun (Chortí) or the phases of the moon (Quiché). These calendars realized the relations between the order of time and space. The regulated and identical orientations of the axes of buildings in the horizon reference-system realized the relations to sunrise and sunset and to the calendar, e.g. the "16-17° family" with Teotihuacan and the other groups, the deviations of around 7°, 11.5°, 20.5° and the solstitial direction 25°, with 4.5° between each other. There are the plans of settlements and fields in the great basins of Central Mexico with the same orientations.

The planning of settlements, we postulate, was realized by means of radial sightlines, beginning in an important center as in the Sun-Pyramid of Teotihuacan. The radians could have directions measured or plotted in the circle of 80 units of 4.5°. The centers of Maya cities also seem to be planned with this method as Copan, Tikal and Uxmal.

Mona Spangler Phillips

Doctoral Candidate; Case Western Reserve University, Cleveland, OH

MEGALITHIC GEOMETRY

We try to learn something about the intellectual development of our remote ancestors by studying the artifacts they left. The megalithic monuments of the British Isles have been a particular focus of such studies. Professor Alexander Thom, having analyzed surveys he has made, concludes that the builders had aligned some of the monuments astronomically; that standard measuring units had been used; and that the shapes of some stone rings followed geometric schemes. In this paper, I verify and modify Thom's geometric hypothesis.

There is one type of monument for which Thom does not propose geometric planning: the type which consists of concentric rings of stones. I find that all sites of this type which are surveyed in <u>Megalithic Rings</u> (A. Thom and Aubrey Burl, B.A.R. British Series 81, 1980, Oxford) can be modeled by variations of one basic geometric motif. I find that this same motif provides alternatives to Thom's proposed geometric schemes for the "flattened circles." Moreover, in those cases where flattened circles are enclosed in concentric rings, my scheme replicates both flattened circle and enclosing ring. I theorize that the motif I illustrate was traditionally used in planning monumental sites and that it had symbolic significance in neolithic Britain.

There is a tendency to project our own values into the past. Several scholars find megalithic geometry anachronistic because it has been taken as implying objective mathematical science. Considering the special nature of the motif in question, I suggest a different interpretation. I propose that certain geometric relationships were noticed during craft work, were associated with fundamental concerns such as fertility or regeneration, and were assigned symbolic roles in cult or religious observances. Although such use of geometry might not be considered "scientific," it indicates an intellectual development which is a necessary preliminary to scientific thinking: the ability to conceptualize and process abstract relationships.

Brent Berlin

University of California, Berkeley

On the non-utilitarian bases of ethnobiological classification

The field of ethnobiology has traditionally focused on the study of the use of plants and animals by non-literate populations. Since the 1950s, interest in the field has expanded in scope, growing to encompass not only the utilitarian factors surrounding the cultural significance of living things for human populations, but more generally to include what the ethnobiologist Ralph Bulmer has suggested to be the "...study of human conceptualization and classification of plants and animals, and knowledge and belief concerning biological processes" *(Social Science Information*, Vol. 2, 1974, p. 9).

An enduring issue in ethnobiology concerns the bases of ethnobiological knowledge. Two competing explanations have been suggested in the literature. One, associated with Claude Lévi-Strauss, can be called *intellectualist* in that "...the universe is an object of thought at least as much as it is a means of satisfying needs" (*The Savage Mind*, 1966, p. 2). A second position, referred to by Hays as *utilitarian-adaptationist*, argues that ethnobiological classification is fundamentally elaborated for the sole purpose of assisting human populations in adjusting to their particular habitats by recognizing and ultimately classifying plants and animals that have practical consequences for the populations involved. (See Hays, *Journal of Ethnobiology*, Vol. 2, 1982, pp. 89-94). The most vocal proponents of this second view are Hunn (*American Anthropologist* Vol. 81, 1983), Hunn and Randall (*American Ethnologist* Vol 12, 1984), and Morris (*Journal of Ethnobiology* Vol 4, 1984).

The present paper provides evidence supporting an essentially intellectualist view on the nature of ethnobiological classification, arguing that the objective discontinuities in nature that form the basic units of ethnobiological taxonomies are unambiguously perceptually based. While not denying the importance of practical concerns for indigenous peoples in their day-to-day relationship with the biological world, field work from a number of investigators indicate that much of what is known of nature cannot be shown to have direct or indirect utilitarian significance. Data from indigenous populations of southern Mexico and the Upper Amazon are examined, as well as experimental results of psychological research with naive American college students. These materials show functionalist arguments to be inadequate to account for the obtained findings.

Eloy Rodriguez, Richard Wrangham and G.H.N. Towers
Univ. of Claifornia, Irvine, Univ. of Michigan, Ann Arbor and
Univ. British Columbia, Vancouver Canada

SIGNIFICANCE OF MEDICINAL PLANT SELECTION BY WILD CHIMPANZEES AND MAN

Recent studies by Wrangham, Rodriguez, Nishida and Towers have established that wild chimpanzees in Tanzania, Africa are eating and swallowing plants in a very unusual manner suggesting possible pharmacological effects. Chemical analysis of one African species of the sunflower family has resulted in the isolation and identification of a potent antibiotic and possible anhelmintic. The plant, <u>Aspilia</u> is also used by Africans for curing intestinal worms. The significance of plant selection by wild apes and man will be discussed.

Dorothy Hosler

Postdoctoral Fellow: Center for Materials Research in Archaeology and Ethnology. Massachusetts Institute of Technology

CULTURAL ATTITUDES AND THE METALLURGY OF ANCIENT WEST MEXICO

Archaeological studies of ancient metallurgies indicate that cultural attitudes toward metal played a primary role in determining the characteristics of those technologies. The metalworkers of ancient West Mexico used various metals and alloys to fabricate objects that functioned in the elite or symbolic and sacred realm of culture. Metal objects included those that were exclusively ritual or symbolic and others that were tools but used in ritual activities. Bronze- the alloy of copper and tin- was the most widely used alloy in West Mexico. Among objects made of bronze were large elaborately crafted ceremonial tweezers worn as items of personal adornment by priests to communicate power and elite status. Small tweezers, of the same material were implements used presumably for depilatory functions. Bronze was also used to fabricate axes used for cutting wood, an activity with symbolic and religous significance in West Mexico. In each case, the alloying element, tin, was added to copper in concentrations appropriate to the objects' design and function. Status items contained tin in high concentrations, making the color of the object golden and visually dramatic. The tools contained tin in lesser concentrations, sufficient to make the metal hard, when cold worked, a necessary attribute for a cutting tool. The West Mexican metalworkers understood and manipulated two properties of the alloy-color on one hand, and its ability to work harden on the other. While all aspects of West Mexican metallurgy reflect the clear understanding of the properties of metals and alloys, the overall configuration of the technology reflects a perception of the material that was cultural: metal was a material properly used for objects that were symbols in the elite and religous or ceremonial realm of culture, or for tools to carry out activities in that same realm of life.

Peter F. Lingwood

Independant Researcher and Winston Churchill Travelling Fellowship
8 Sorrento Way, Darfield, Barnsley, South Yorkshire.

THE PACIFIC AND THE MAKING OF AN ADMIRAL - EDWARD BELCHER (1790-1877)
AND THE VOYAGE OF HMS 'BLOSSOM'

Many different routes led to the rank of Admiral in the British Navy
during the nineteenth century but perhaps none so bizarre to today's
Senior Service than that of Sir Edward Belcher, KCB (1790-1877) who,
amongst other attributes, presided over the amassing of significant
natural history collections.

There is no published evidence to suggest that Belcher held an early
interest in natural history, however, the three year voyage of HMS
'Blossom' to the Pacific (1825-1828), under the command of Frederick
William Beechey (1796-1856), exposed him to a wide variety of
natural history disciplines in some of which he took an active
interest. Indeed, he went so far as to complain in his journal that
insufficient opportunities were provided for natural history
collecting even when the voyage was instigated for hydrographic
surveying as support for an attempt to discover the north-west
passage.

Belcher showed considerable interest, during the voyage, in geology,
but this waned later in life, and in conchology, which resulted
during his lifetime in a considerable personal collection of shells
which required, in 1877, two days to auction. It is also noteworthy
that subsequent voyages to the Pacific under his command, i.e.
HMS 'Sulphur' (1835-1842) and HMS 'Samarang' (1842-47) returned with
significant collections, especially of shells.

(The significance of his personal contribution to these collections
has been assessed elsewhere - Lingwood, P. F., 1985. Admiral Sir
Edward Belcher (1790-1877): natural history catalyst or catastrophe.
From Linnaeus to Darwin Commentaries on the history of biology and
geology. Society for the History of Natural History, London).

It is difficult to assess exactly what, if any, Belcher's interest in
natural history had on his promotion. He was certainly an industri-
ous and talented surveyor with a happy ability to seize upon
opportunities for quasi-military and political adventures for which
he was duly honoured. His career, however, prospered during a
period when there was a considerable interest in other continents and
in their peculiar productions both by polite society and by the Navy
in general, and his superiors in particular. His interest certainly
appears to have done him no harm.

The initial voyage with Beechey appears, therefore, to have planted
a seed which Blossomed and flourished throughout, and may have
helped to shape, Belcher's eventful career.

Alan Frost,
Department of History,
La trobe University,
Bundoora, Victoria 3083.

Science for Political Purposes: The European Nations' Explorations of the Pacific Ocean, 1764-1806

In their public aspects all the great voyages of Pacific exploration from Byron's and Wallis's to Flinders's and Baudin's were profoundly scientific ventures, which separately and together added greatly to "the mass of our knowledge."

Yet each voyage had other, less publicized purposes, which involved establishing rights to possess territory and control trade, or countering the steps taken by others to gain territorial, strategic, or economic advantage. I shall discuss these concealed aspects of the voyages in the contexts of European politics and imperial rivalry in the second half of the eighteenth century.

Kenneth J. Carpenter

Professor of Nutrition, University of California, Berkeley

The Pacific as a Testing Ground for Theories of Scurvy

One great problem for early navigators in long voyages across the Pacific was scurvy amongst the crew. This was attributed, most commonly, to cold sea air which: "produces with its subtlety and coolness some corruption of bad humor in persons worn out and fatigued." They believed the corruption came from blockage of perspiration considered the main route for excreting toxic metabolites. Cleanliness, protective clothing, rest and regular exercise were all thought important for maintaining health.

The Anson expedition (1747-49) first had scurvy after rounding Cape Horn in bad weather. After recovering on Juan Fernandez, they had a second, serious outbreak when in good weather off the Mexican coast, "which the surgeon could not account for." The naval issue at that time for the treatment of the disease at sea was elixir of vitriol (flavored sulphuric acid).

For Cook's three expeditions into the Pacific (1768-80), he was given a range of anti-scorbutics to test. Foremost was <u>malt</u> (dried, sprouted barley). David McBride recommended this on the ground that, infused with water, it made a drink (wort) which fermented rapidly to give 'fixed air' and that this would reconstitute (or re-fuse) putrefying tissues on the point of decomposition in scurvy. Since putrefying meat gave off a gas, it seemed logical that the 'fixed air' in them had formed their cement.

Cook kept his expeditions free from scurvy. He took hygenic precautions, and also made frequent stops at islands where he made his men eat fresh green vegetables of all sorts. He also got them to eat sauerkraut at sea. He was awarded a Royal Society medal for his success, which the President attributed primarily to the use of malt. Banks, with the first expedition as a scientist, himself developed mild scurvy although drinking wort, and cured it with a private stock of lemon juice, but his diary was not published in that period. Even he wrote later in favor of McBride's 'ingenious treatise' with its attractive theoretical basis.

Modern analyses show a negligible value for wort, which confirm the results of later tests (1785-1800) in the British Navy in the Atlantic. The theory also ceased to be attractive when it was realised that the same gas (CO_2) was produced in respiration and fermentation. Success in the prevention of scurvy in large fleets came only with the use of lemon juice. It is now thought that Cook's reports "delayed rather than hastened the introduction of the true cure for scurvy."

Miranda Hughes

Ph.D Student   University of Melbourne

The Société des Observateurs de l'Homme and the Development of Anthropology

This paper focuses on one particular society, the Société des Observateurs de l'Homme (1799-1805), and its influence on the nascent science of Anthropology. Its membership was highly representative of Parisian intellectual life at the turn of the Nineteenth Century. I argue that the work of this Society reflects the fundamental reorientation in perspective which resulted in the adoption of a scientific mode of discourse as the primary means by which the problems of human nature were discussed. The Society was instrumental in sending the first field trip of anthropology which was the voyage of the French scientific expedition, captained by Nicholas Baudin, to the Southern Hemisphere. This voyage spent the summer of 1802 in Tasmania, Australia, observing the natives. Many papers of physical anthropology and cultural ethnology were presented to the Société to aid the scientists in their observations. These papers present an analysis of the methods and apparatus they thought appropriate for anthropology. In them, explicit comparisons are made with the methods of the physical sciences. This was the first voyage of exploration to have had the 'scientific' aim of the study of Man written into the official instructions. The resultant reports, accounts and illustrations of the voyage form an unique source of information about the Tasmanian Aborigines. These documents, inevitably and understandably, exhibit some superficial Eurocentric interpretations of the rituals and customs of the Tasmanians. However, we can find the intention and aim of evaluating each society according to its own internal standards, rather than by prevailing European customs, written into the anthropological instructions and strived after by the field scientists. They therefore provide many interesting insights for an historical sociology of science.

Barry Butcher

Tutor, Deakin University.   Post Graduate student (Ph.D), University of
Melbourne

AUSTRALIAN CORRESPONDENTS OF CHARLES DARWIN: A CASE OF 'JUDICIOUS SELECTION'?

The use that Darwin made of material gathered by correspondents from around the world is well known. Among these often obscure workers in the cause of evolutionary theory were many from Australia; men who wrote from first-hand experience of the continent's flora and fauna and/or its "savage and degraded" Aboriginal inhabitants. Their reports proved to be an invaluable source of information for Darwin, both in formulating and in buttressing his theories of Natural and Sexual Selection. Some of this information will be discussed in detail; in particular, that resulting from the circulation of his <u>Queries about Expression</u> of 1867, the responses to which formed the backbone of <u>The Expression of the Emotions in Man and Animals</u> of 1872.

An analysis of the <u>Queries</u> and the Australian responses to them reveals, firstly, that the material used by Darwin was 'selected' from the information received at Down, in order to highlight key points in his argument; and, secondly, that this selection increased a bias, probably unconscious, already inherent in Darwin's use of a 'questionnaire' procedure.

Roy MacLeod

Professor, Department of History, Sydney University

Imperial Reflections in the Southern Seas: The Funafuti Expeditions, 1896-1904.

Traditionally, the history of European science in the Pacific is inseparable from the history of European great power rivalries. Science accompanied the flag. Even so, the history of Pacific exploration and discovery in the 19th century set new standards for international cooperation among men of science, sometimes pursuing national interests in the quest for fundamental knowledge.

By its nature, much of this activity was undertaken and directed from metropolitan centres of scientific learning. By the 1890s, however, in a Pacific dotted with European outposts, and witnessing the presence of new colonial interests, the scene was set for new forms of "imperial science." This paper will discuss the changing nature of scientific imperialism in the Pacific, with reference to the British and Australian expeditions to the atoll of Funafuti in 1896 and 1897. These expeditions, sent to test Darwin's theory of coral reef formation, began a new stage in geological investigation and in the history of Australian scientific endeavour. This account will reflect upon Imperial tensions and ambivalences that, in the wake of Funafuti, were later to help shape British and Australian scientific activity in the Pacific, the Southern Ocean and Antarctica.

R.W. Home, University of Melbourne, Melbourne, Australia

Masao Watanabe, Tokyo Denki University, Tokyo, Japan

PHYSICAL SCIENCE IN AUSTRALIA AND JAPAN TO 1914:   A COMPARISON

Physics first became established in Australia and Japan at the same period, during the final quarter of the 19th and the first years of the 20th century. In this paper we offer a comparative study of the processes by which this happened in these two developing countries on the Pacific rim. We show that, despite the great cultural differences that existed, and that might have been expected to have been a source of major differences in national receptiveness to the new science, there were in fact many parallels between the patterns of development in the two cases. Identifying these enables us to draw attention to a number of significant features of the physics discipline more generally at this period. Such differences as emerge in the early history of physics in the two countries seem to have arisen more from the different political situations that prevailed than from anything else; in particular, they reflect the fact that Australia was a part of the British Empire while Japan was an independent political power.

Robert Randolph and John Bardach

East-West Resource Systems Institute, Honolulu, Hawaii

SOVIET SCIENCE IN THE PACIFIC: THE CASE OF MARINE BIOLOGY

Pacific Ocean research by both Imperial Russian and Soviet scientists has been uniquely conditioned by geographical, economic, political, and other circumstances. Focusing on the example of marine biology, this paper highlights noteworthy aspects of Soviet science in the Pacific, including peculiarities of institutional development, research emphases, and both theoretical and practical conclusions. It is shown, for instance, that the Soviet worldview led to a relatively early and explicit realization of the need for husbandry and management of the hydrosphere, and to specific proposals as diverse as nuclear-powered artificial upwellings, whale farms, and chemical fertilization of the open ocean. While not all such proposals are likely to be implemented, others have been already, such as the recent creation of the USSR's first marine park, near Vladivostok.

Chen Yanhang and Yang Qiuping

Gimei Navigation Institute

A new verification of the type of Zhenghe's treasure vessel

850 years ago, Zhenghe, the well known navigator of the Ming Dynasty, heading a large fleet with a crew of 20-30 thousand, sailed to the Western Ocean for the first time in Chinese history. In the next 28 years (1405-1433) Zhenghe and his men made seven seccessive voyages to the Western Ocean and visited more than 30 countries in the south-east Pacific and along the coast of the Indian Ocean. They went as far as the east coast of Africa, more than 6,000 miles away. Zhenghe's expeditions not only brought about technical and cultural exchanges, but forged friendly relations between China and those countries, and thus promoted the flourishing of China's marine "Silk and Porcelain Road."

The replica of Zhenghe's treasure vessel, now on display in the Museum of Chinese History, Beijing, was designed and made after the pattern of Nanjing shachuan (large junk). The authors of the present article, in the light of the historical documents and the latest relevant materials, have re-verified the type of Zhenghe's treasure vessel and come to the conclusion that it was of the Fujian type. 44.4 zhang (120 meters) in length and 18 zhang (50 meters) in breadth, the treasure vessel was magnificent and luxurious with two dragon eyes sculpted on both bows and the bottom painted white. The ship, which was fitted with 9 masts and 12 sails, weighed about 7 thousand tons.

Harry N. Scheiber,

Prof. of Law and History, Univ. of California, Berkeley

WILBERT CHAPMAN AND THE REVOLUTION IN U.S. PACIFIC OCEAN
SCIENCES AND POLICY, 1945-60

The post-World War II period witnessed profound changes in
U.S. ocean policy--changes that reflected transformations in
the technology of fisheries, in the ocean sciences, and in
international law and diplomacy. This paper considers the role
of the biologist Wilbert Chapman, a major actor in the
policy process in the United States, in these transformations.
Chapman was one of the principal architects of major new
initiatives in Pacific fisheries science. He was a principal
organizer of the California state Marine Research Committee
research project for study of the sardine depletion in the
California Current. He also played a key role in organization
of the federally sponsored POFI (Pacific Oceanic Fisheries
Investigation), based in Hawaii. Both these new scientific
undertakings contributed toward the integration of fishery
biology with traditional oceanography in the 1940s and early
1950s. Concomitant advances in fishery science were exploited
also in the Inter-American Tropical Tuna Commission studies
of the 1950s, conducted by an agency that Chapman helped to
establish by virtue of his role as a U.S. diplomat (1948-52).
Both in his State Department post and in the private sector
after 1952, Chapman made contributions of enduring importance
in the development of American ocean policy and of Law of the
Sea, continuing all the while to promote organized science
and the continuing transformation of commercial fisheries
research in relation to studies in biology and oceanography.
Among his principal achievements, moreover, was the fusion of
commercial, academic, and governmental scientific efforts in
a variety of research arenas in ocean sciences.

Garry J. Tee

Senior Lecturer, University of Auckland, Auckland, New Zealand.

MATHEMATICAL SCIENCE IN NEW ZEALAND.

The Maori of New Zealand developed an advanced Neolithic culture, with strictly decimal numeration. New Zealand was added to the British Empire in 1840, & European culture (including mathematics) was then introduced rapidly. The University of Otago was founded in 1869, & in 1874 it was absorbed into the University of New Zealand. In 1961 that University of New Zealand dissolved into 6 universities.

Libraries in New Zealand hold many early mathematical books, including a mediaeval Arabic manuscript of Euclid's *Elements*, & many early editions of Newton & Halley. There are some manuscripts of many mathematicians, including William Rowan Hamilton (1805-1865); & there are very many relics of Charles Babbage (1791-1871).

Ernest Rutherford (1871-1937) studied at Canterbury University College (1890-1895), where he gained expertise in mathematics which enabled him to discover the atomic nucleus, in 1911. In the early 20th century D. M. Y. Somerville (at Victoria University of Wellington) & R. J. T. Bell (at the University of Otago) wrote mathematics texts which are still being used. Leslie John Comrie (1893-1950) graduated at Auckland University College in 1916, & he became the leader in scientific computation throughout the second quarter of the 20th century. Alexander Craig Aitken (1895-1967) graduated at the University of Otago in 1919, then went to the University of Edinburgh in 1923. He became a very distinguished mathematician, in algebra, numerical analysis & statistics. The eminent geometer Henry George Forder (1889-1981) came from England in 1934, to become Professor of Mathematics at Auckland University College.

Professor Roy Kerr (University of Canterbury) shewed in 1963 that black holes could rotate.

Philip F. Rehbock

Assoc. Prof. of History and General Science, University of Hawaii

ORGANIZING PACIFIC SCIENCE: ORIGINS AND EVOLUTION OF THE PACIFIC SCIENCE ASSOCIATION

In reaction to the tragedy of World War I, a wave of internationalist sentiment was propagated, touching many lands and stimulating the establishment of numerous institutions for the promotion of peaceful solutions to global problems. This wave soon reached the U.S. Territory of Hawaii, where several international organizations sprang into being. Their scope was not global but rather Pacific-wide, or "Pan-Pacific"--the common phrase of that era. Hawaii had begun to see herself as the commercial entrepôt and cultural focus for Pan-Pacific activities. In this context it is not surprising that the first effort to create an institution for coordinating and communicating the results of Pacific-based scientific research should come from Hawaii. And no institution was a more likely candidate for promotion of this goal than the Bernice P. Bishop Museum of Honolulu, which had by then been Hawaii's symbol of scientific endeavor for 30 years. What is surprising is that the product of this effort, the Pacific (originally Pan-Pacific) Science Association grew up, survived World War II, and seems now, at age 65, to be in its prime, while the great bulk of the progeny of the 1920s' internationalism have disappeared.

Two figures stand out as the creators of the PSA: Herbert E. Gregory and Alexander Hume Ford. Gregory was Silliman Professor of Geology at Yale but had arrived in Honolulu in 1919 to become director of the Bishop Museum. The following year he presided over the 1st Pan-Pacific Congress which assembled 93 scientists from nine countries around the Pacific at the hub in Honolulu for 18 days of papers, fellowship, and planning. Gregory continued to guide the fortunes of the PSA for the next 30 years. Ford was a Honolulu publicist, outdoorsman, and founder of the Pan-Pacific Union, a Honolulu organization operating since 1911 to improve awareness among Pacific peoples of their common problems and stimulate cooperation toward solutions. The Pan-Pacific Union issued the initial call for the 1st Congress and supervised the funding made available by the territorial legislature. Subsequent congresses were held at three- and later four-year intervals, with an understandable hiatus during the World War II years. Hosting countries have taken the congresses from Vancouver and Auckland to Bangkok and Khabarovsk. An analysis of papers presented at the fifteen congresses and five intercongresses held to date provides one important index for establishing the meaning of the term "Pacific Science," its evolution in the 20th Century, and the extent to which Western perspectives have been dominant.

Alan E. Leviton, California Academy of Sciences, San Francisco, CA

Michele L. Aldrich, American Association for the Advancement of Science, Washington, DC

THE CALIFORNIA ACADEMY OF SCIENCES, 1853-1906: Its First Half Century.

California, following several years of conflict between Mexican authorities and the growing resident "American" population, became the 31st state of the United States in 1850. Three years later, in the city of San Francisco, seven amateur naturalists met to found the California Academy of Natural Sciences. The founders observed that "Scientific associations have been organized in many of the older states whose investigations and labors have brought to light many of the previously hidden mysteries of nature and have contributed immensely to the progress of the age in the practical application of the natural laws to the purposes of agriculture, commerce and the useful arts." Five of the seven founders were medical practitioners, one a notary public, and one, superintendent of public schools.

Academy members met weekly, and at these meetings read papers based on their own studies of plants, animals, and geology. Albert Kellogg, William Orville Ayres and John Boardman Trask were notably active. Within 18 months of its founding, the fledgling Academy published volume one of its Proceedings, the first scientific journal of its kind in the American West. However, unofficial published proceedings of the Academy's weekly meetings appeared even earlier, in a non-sectarian religious weekly newspaper, The Pacific. Scarcely nine months after its founding, the Academy also began acquiring specimens of local and exotic plants and animals, the beginnings of the most celebrated "cabinet" (museum) in western North America.

The association rapidly emerged as an intellectual oasis in a state mostly concerned with more practical matters, such as the decline of its gold mining industry. It attracted the attention of Eastern and European scientists: Joseph Henry and Spencer Fullerton Baird of the Smithsonian Institution, Louis Agassiz, the maverick Hungarian adventurer John Xantus, Matthew Fontaine Maury, Thomas H. Huxley, and others who contributed advice, library materials, specimens, and papers for publication.

With the founding of the University of California in Berkeley in the 1860s and three decades later Stanford University, the scientist membership and intellectual consciousness of the Academy expanded. On the 18th of April, 1906 the city of San Francisco was devastated by an earthquake and fire, which engulfed the Academy's building as well. Nearly all collections and records were lost, but in November, 1906, the schooner "Academy" returned from an 18-month voyage of exploration to the Galapagos Archipelago, to open a new chapter in the history of the organization.

Elizabeth Newland,

Ph.D. Student, Tutor, Department of History & Philosophy of Science,
University of Wollongong

FROM SIR JOSEPH BANKS TO THE AUSTRALASIAN ASSOCIATION FOR THE ADVANCMENT OF SCIENCE. THE TRANSITION FROM 'DEPENDENT SCIENCE TO 'INDEPENDENT SCIENCE'. AUSTRALIA 1788-1888

ABSTRACT

Australian science in its colonial period was shaped by several factors. They were:
i) The penal settlement origins of the colony of New South Wales
ii) The structure of Australian Colonial society.
iii) The so-called 'Tyranny of Distance'. The twin problems of Australia's isolation from Britain, the main scientific centre of 19th Century science for the English-speaking world and the vast distances and poor communications within Australia affected the colonial scientist.
iv) The hegemony, as typified by Sir Joseph Banks of British Science and scientific institutions.

This paper examines in detail each of these factors and traces Australian science from the first settlement at Farm Cove (Sydney, New South Wales) in 1788 to the emergence of an independent scientific tradition with the establishment of the first national scientific society - the Australiasian Association for the Advancement of Science - in 1888.

Michel T. Halbouty

Chairman and CEO, Michel T. Halbouty Energy Co., Houston, Texas, USA

HISTORY OF PETROLEUM EXPLORATION IN THE UNITED STATES

Petroleum exploration in the United States, and worldwide, has followed a very logical and economic pattern. It has been conducted in areas offering the greatest reward: areas where the geology is most favorable; where chances are best of finding the most oil and gas cheaply; where petroleum can be produced and marketed at the lowest cost; where markets are the strongest, prices highest, and where stable political and economic conditions can be expected.

From the first discovery of oil from a drilled well at Titusville, Pennsylvania, in 1859 to today, the history of petroleum exploration in the United States has been marked by dramatic discoveries of oil and gas—such as the giant East Texas Field and the Prudhoe Bay discoveries, and with equally dramatic "failures", such as Mukluk—a multi-billion dollar dry hole. Yet regardless of such failures, knowledge is added, theories confirmed or refuted in the unrelenting scientific quest.

From prospecting with witch-hazel twigs, to using conceptually-derived maps of the paleogeomorphology of a region, to using the most sophisticated seismics and computer-generated models, petroleum geoscientists have charted a fascinating course in the records of American innovation. Engineers and technologists have contributed a constantly-improving flow of designs for drilling and production equipment. The petro-professions are constantly creating new concepts in exploration, drilling and production—but probably the most exciting and rewarding are those which have not yet been formulated.

The significance of the United States petroleum industry's record in petroleum exploration lies in the basic belief that the explorer has to use all knowledge available, all of the tools, and have the strength of his convictions to search where others fear to tread.

Stanislaw Czarniecki, Polish Academy of Sciences,
Kraków, Poland

POLISH PETROLEUM EXPLORERS IN EUROPE, ASIA AND AMERICA BEFORE THE REGAINING OF THE INDEPENDENCE OF POLAND.

The beginnings of Polish geology coincide with the decline of Polish state in the second half of the 18th century. For the "Father of Polish geology" Stanisław Staszic /1815/ the new science seemed to be one of possible means to maintain the life of the nation.

In 1853 two druggists from Lwów, I.Łukasiewicz and J.Zeh obtained the patent on crude oil refining method and invented kerosene lamp. Soon, oil become the most important natural resource of the Carpathians.

In both university centers of Galicia, in Kraków and Lwów, there worked several petroleum geologists who were well recognised in the world. J.Grzybowski introduced micropaleontology to stratigraphic investigations of the Carpathians /1897/. W.Teisseyre, while working in Roumania in 1896-1910, with L.Mrazec formulated the theory of diapire tectonics. R.Zuber's work "Flysch and petroleum" /1918/ was the first study on the origin of oil. Before 1918 fifteen Polish geologists were involved in oil exploration in Hungary, Roumania, Italy, Peru, Argentina, Venezuela, Sumatra, Guinea and trans-Caucasus oil fields.

It should be stressed that most Polish petroleum geologists came from Galicia, while only four of them from the trritories under Prussian or Russian occupation. This was because independent and selfgoverning Polish scientific institutions could develope only in the territory included in Austria. For a nation deprived of an independent state the developing of science on its own is more important than to have access to foreign universities, even when their lavel is high.

WANG YANGZHI

PETROLEUM ADMIN IN NORTHERN CHINA

CHINESE OIL EXPLORATION FOR 100 YEARS

The hundred years from the beginning of prospecting in Taiwan Province to the time the output of Chinese crude oil rose to 100,000,000 tons can be divided roughly into four stages. During the first, from 1878 to 1920, a few fields were opened and Taiwan carefully surveyed, primarily under the direction of foreigners. During the second phase, from 1921 to 1948, native Chinese took over expoloration and exploitation, chiefly in Gansu and Xinjiang. Between 1939 and 1948 they yielded 450,000 tons of oil, some 95 per cent of the output of the whole country.
The third stage began with the founding of the People's Republic of China on October 1, 1949. The whole country possessed three small oil fields, eight worn-out drilling machines, and 52 oil wells. The annual output of petroleum was no more than 120,000 tons. The "May 1" model of seismograph went into operation where iron hammers, compasses, tape measures, etc., had been applied before. Russian-made rigs were introduced that could drill a well 3000 meters deep. In 1959, the output of crude oil, 2,758,000 tons, was 23 times that of 1949.
The fourth stage, since 1960, began with the opening of the Daqing Oil Field and has brought in more than 180 fields. Since 1978, the annual output of oil has exceeded 100,000,000 tons. The reasons that our petroleum industry triumphantly developed from weak to strong and from small to large are:
1. Prospecting. Geologists Pan Zhongxiang and Huang Jiqing proved, against established opinion, that land facies (which make up most of the Cenozoic and Mesozoic strata in China) could form an oil field with industrial value. The source bed of Daqing is land facies.
2. Prospecting technology. After adoption of the "May 1" seismograph, improvement came rapidly: tape seismographs, number seismographs, electronic computers, and deep drilling rigs.
3. The socialist system. It paves the way for the development of petroleum prospecting.

Qi Shuqin

Editor, Shanxi Bureau of Seismology, Taiyuan, China

EARTHQUAKE SCIENCE IN CHINA'S HISTORICAL DOCUMENTS BEFORE THE 20th CENTURY

Through induction, arrangement and classification of China's historical earthquake science documents before 20th century this paper discussed certain development and achievement in China's earthquake science, with an emphasis lied on the 19th century.
1. The rich storage of earthquake science information in China's historical documents has a close relation with China's long history and frequent earthquake occurance. The great amount of information which is of value to the earthquake science research contained in various local historical documents is a sound proof. A deep-going research, arrangement and utilization of this materialis an important subject in the study of earthquake history.
2. The high tide of earthquake occurance provides a valuable condition and opportunity for the research experimentations and hitherforth a way of achieving new results in this field can be opened up.
3. The rich content of historical documents of earthquake shows a ever-deeper trace in the understanding and the exploration of earthquakes, in which can be found the scientific summing up of indications of earthquakes, and its forecasts , earthquake disasters and its social influence , earthquake projects and earthquakes precautions and anti-earthquakes etc. The great amount of unusual earthquake records and the vivid disoription of every earthquake accident shows that there are indications of earthquakes and there are laws which can be discovered, which are the important scientific basis for earthquakes forecasts. And the summery of the experience of earthquake projects and the earthquake relief work are the historical lessons in making anti-strong-earthquake plans.
4. China's earthquake science in 19th century has made a great stride in the current of development of modern science and technology and on the basis of the summery and the inheritance of historical achievements. This is of great significance in inheriting the past and ushering the future in the development of 20th century China's modern earthquake science.

Susan Schultz

Park Historian, National Park Service, Olympic National Park

FROM SCENERY TO SCIENCE: THE DEVELOPMENT OF GEOLOGIC INTERPRETATION IN THE NATIONAL PARKS

From their beginning, American national parks have served as centers of geological research and public education. The first parks were set aside in recognition of their spectacular, rugged scenery-- a consequence of their geologic history. Yosemite and Yellowstone, with their vivid, awe-inspiring geologic phenomena, represented the American park ideal many years before there was a coherent, systematic administration of national parks. The interpretation of geology in the first national parks was carried on by soldiers, stagecoach drivers, and inn keepers. However, as the system of national parks expanded, it was recognized that the parks, individually and collectively, represented nothing less than "the geologic sequence of America's making." With the establishment of the National Park Service in 1916, its first director, Stephen Mather, emphasized that one of the chief functions of the national parks would be to serve educational purposes. Formal interpretive programs began in the national parks in 1920, and geologists were among the first seasonal rangers who interpreted the natural history of the parks to visitors. This paper examines the evolution of geologic interpretation in several national parks in the American West.

Martin J S Rudwick

Trinity College, University of Cambridge, England

ARENAS OF GEOLOGICAL DEBATE IN EARLY 19TH-CENTURY EUROPE

The historiography of Karl von Zittel's "heroische Zeitalter der Geologie" (Geschichte der Geologie und Paläontologie, 1899), combined with a nationalistic bias that Zittel rejected, continues to dominate current research on the history of the earth sciences. This should be replaced by an approach which (a) examines the interactions between leading scientists and lesser figures, and among the leaders themselves; (b) attends to the international dimension of this process of exchange; and (c) studies the whole range of scientists' activities and not only their formal publications. In this paper, such an approach will be illustrated briefly by taking a single sample year from the early 19th century, and by reviewing one form of scientific activity within it. The chosen year is 1835 (i.e. exactly a century and a half ago); the chosen activity is that of face-to-face discussion among groups of geologists - arguably the most important means of geological education at that period. The paper will focus (a) on the membership and meetings of the two premier geological societies of the time (Geological Society of London; Société Géologique de France) and the geological Sections of the two major peripatetic scientific organisations (Gesellschaft Deutscher Naturforscher und Ärzte; British Association for the Advancement of Science); and (b) on the character of the geological problems debated in those arenas in 1835.

Gregory Benford

Physics Dept., Univ. Calif., Irvine

THE SCIENTIST IN FICTION

Modern literature reflects only dimly the transforming power of the scientist. Since the ambivalences of Frankenstein, writers have fixed on the 'technohero' aspect, looking at larger social issues without understanding scientists as a distinctive type of personality. The Strangelovian image further distanced fiction from real scientists by stressing moral issues and the insanities of ungoverned technology. Few have tried to understand scientists as personalities at a unique intersection of social and intellectual forces, and instead often oversimplify them as romantic idealists, like artists. A few works, some notably science fictional, have addressed this problem. A primary difficulty is that scientists are not trained as writers, and the great range of scientific "types" remains largely unexplored.

Robert M. Philmus

Professor of English, Concordia University (Montreal)

Olaf Stapledon's Tragi-Cosmic Vision

Olaf Stapledon was evidently indifferent to matters of genre; as a writer, his commitment was to the propagating of ideas. This, however, makes his major works--from <u>Last and First Men</u> (1930) to <u>Sirius</u> (1944)--all the more significant: "happening" to be science fiction, they confirm that genre's vocation for ideas.

The ideas that Stapledon most often concerns himself with are of the cosmic variety. On the basis of what might be called a post-Darwinian understanding of evolution, he elaborates a "tragi-cosmic" vision of humankind's fitful "spiritual" progress on a journey to which time must eventually put an end but which otherwise has no destination, no finality about it. Moreover, his fictions have a kind of participatory intent. Rather than being objects for passive consumption, they are designed to awaken the reader to possibilities of consciousness which apparently have their locus elsewhere or elsewhen, but really reside in her or his own mind.

George Slusser

Curator/Adjunct Professor   UC Riverside

Sciences of the Mind in French Science Fiction

The new <u>cogito</u> for French science fiction seems to be "I think, therefore the universe is." In relation to the physical universe, this may seem reductionist, may reflect a desire to mitigate the terrors of Descartes's <u>res extensa</u> by making mind coterminous with world. I see something quite different. For the past twenty years French science fiction has been fascinated with patterns which, rather than reductive, I call introspective. Nuclear holocaust for instance, an obsessive theme, becomes in this literature a creator of dreamscapes, a means not only of isolating individuals but of forcing them back into their own minds in order to seek solutions to external conditions. The result of this introspective focus may be paranoia or amnesia--disorders of the mind--or it may be renewed exploration of that mind as place. This latter is true of the French version of time travel, which is specifically seen as mind travel, and as chronolysis the scientific investigation of time lines, of worlds created by and within the human mind.

In this fiction then, uncertainty is often relocated within the confines of the mind, and its gaps and "walls" have (as metaphor and as reality) become one with those of the physical world itself. The possibility of scientific investigation remains, but its field is now the mind. The frontier here lies in unexplored neural and synaptic links, not as in the American model in vast outer space. And breakthrough, where still possible, is now through the needle's eye of these altered states of the rational self. The vehicle of exploration, finally, is not the "hard" sciences--those that reengineer the external universe to fit man and create artificial intelligences to extend his reach--but the sciences of the mind: neurochemistry, the stimulation of expanded consciousness.

Positing that the central metaphor for world in this literature is not (as in American SF) the extended body but rather the expanded mind, this paper examines the fictional working out of such mind explorations in works like Philippe Curval's <u>En souvenir du futur</u>, Christian-Yves Lhostis's <u>Tous ces pas vers le jaune</u>, and especially Michel Jeury's masterpiece <u>Le temps incertain</u>. This fiction is built upon creative tensions between restrictive brain and expansive mind, and these tensions must be seen in relation to, and reaction to, recent scientific attitudes toward the sciences of the mind in France. This focus allows me, finally, to contrast French SF with American works which reflect this nation's fascination with the metaphors of the expansive sciences of nature: astronomy and biology, exploration of Pascal's infinitely large and infinitely small which tends to ignore the central term of his postulation, man thinking, the human mind as process and place.

Mark B. Adams

University of Pennsylvania, Philadelphia, USA

EUGENIC VISIONS (1920–45)

During the interwar period, new prospects for the management of future human evolution excited a group of biologists, philosophers, and novelists. Born around 1890 and influenced by the visions of the early H. G. Wells, these men saw startling possibilities in the "new zoology" and shared J. B. S. Haldane's conviction that "the biologist is the most romantic figure on earth at the present day."

In "Daedalus" (1924), Haldane envisioned a future in which humanity had been saved and transformed by eugenic "ectogenesis"—in vitro fertilization and development of human eggs. "Possible Worlds" (1927) foresaw the genetic molding of humans to suit new social structures and alien planetary environments. Haldane's essays evoked critical responses from B. Russell (Icarus 1924) and A. Huxley (Brave New World 1932), and inspired a sweeping "myth for the industrial age" by philosopher O. Stapledon (Last and First Men 1930), against which C.S. Lewis wrote his Perelandra trilogy. While Haldane thought ectogenesis might become technically feasible and socially acceptable late in the century, others of his generation found sperm more exciting and saw no need to wait. Technically, new work on artificial insemination in livestock suggested that a powerful eugenic technique was already in hand; socially, the USSR seemed eugenically promising, since "no doubt the Russian people has proved an ideal subject for social experimentation" (Haldane 1927).

In 1929, drawing on breeding work by I. Ivanov, Soviet geneticist A. Serebrovsky proposed that thousands of women be artificially inseminated with the sperm of eugenically selected men, noting that each might father 10-50,000 children. (Later Brewer would name this procedure "eutelegenesis".) Ridiculed 1929-32, Serebrovsky abandoned his proposal. But it was resurrected by American geneticist H.J. Muller who worked in the USSR 1933-37. Muller incorporated Serebrovsky's proposal in his 1935 book Out of the Night. In May 1936, impatient for the future and aware of Stalin's power, Muller sent him the book and a letter urging a trial human breeding program. Within months, Muller's former students Levit and Agol were arrested, Levit's institute closed, and Lysenkoists had launched their first major attack on geneticists (December 1936), castigating their wild eugenic schemes. Disheartened, Muller left Russia in early 1937.

Postwar revalations of Nazi practices deprived visionary eugenics of much of its public support, and two decades of discussion largely disappeared from the scientific and literary memory—except in science fiction. Perhaps the best novel of the best science fiction writer of the "golden age"—Beyond This Horizon (1942/48) by R. Heinlein—portrayed a worldwide eugenic society of the near future which reconciled eutelegenesis with individualism, voluntarism, and personal freedom. Like the best science fiction of the period, Heinlein's work mobilized the scientific, philosophical, and social legacy of the previous generation to explore the social implications of science for the human future.

James H. Bunn

Professor of English and Vice Provost, SUNY at Buffalo

TOWARD A UNIFYING THEORY OF METAPHOR AND MODEL FOR LITERATURE AND SCIENCE

A first step in the search for a unified grammar of symbolic representation might well begin with the common structure of model and metaphor. That which unifies them is a logical crux within all signs. The structure of any sign is compounded by a logical disjunction between the scientific question "What is it made of?" and the ethical question "What is it good for?" By way of John Locke and others I show that the combination of these logically different categories of intent into a unified sign amounts to a fundamental 'category mistake' that binds signs, symbols, metaphors and models into a paradoxical category called 'fictions.' What makes a common theory of literature and science possible therefore is a logical split within all sign language, including codes in nature, and not a logical split between literature and science. Models and metaphors perpetuate that category mistake by way of a secondary turn or trope that I characterize as fictive. So I end with a discussion that defines fictions, inelegantly but accurately, as second-order category mistakes.

Dr Elinor Shaffer

Reader in English and Comparative Literature, University of East Anglia, Norwich, England

THREE MODELS OF METHOD IN THE INVESTIGATION OF THE RELATIONS OF LITERATURE AND SCIENCE

The three models of method I propose to explore are both powerful current models of investigation, and historical phases in the development of the history of science itself. They are the highly evolved contextualism of the present (of which a volume such as The Ferment of Knowledge shows several excellent examples); the phenomenological models represented by several French thinkers; and the Natural-philosophical method, the most radical of the three, in that it works from a critique of the principles of scientific investigation, and attempts to suggest through the inquiry into the relations between literature and science another model for scientific investigation itself. Historically speaking, the contextualism of today might be seen as not different in principle from the view of the Enlightenment that the arts and the sciences could and should be considered in their interrelations and as forming a totality of knowledge to be summed up in an encyclopaedic form. However more sophisticated, however more technical the disciplines have become that make up the circle of the arts and sciences, they can nevertheless be accommodated and accounted for without contradiction in a compendious form. Again historically speaking, the phenomenological model may be seen as a response to a positivist model in which the differences in method were seen as more significant than their similarities; thus the work of Canguilhem, Bachelard, Foucault, and Deleuze, for example, arises from a more exacerbated sense of the difficulties of pursuing any such encyclopaedic model, and proposes several routes to the dissolution of the obstacles for long enough to permit an encounter between the diverse disciplines. Finally, the model of an aesthetic rather than a jurisprudential model for scientific investigation, elaborated at the end of the eighteenth century especially by Schelling, has a modern exponent in the work of Joseph Needham and his collaborators in their detailed account of the way in which a completely different philosophy served in China as a basis for scientific and technological advance. It remains to discuss which of these models is most viable, and in which context.

Stuart Peterfreund

Associate Professor, Department of English,
Northeastern University, Boston, MA

The Rise of Energy in Literature and Science

Since the end of the nineteenth century and the beginning of the twentieth, when the development of field theory, special and general relativity, and quantum mechanics took place, it has been all but impossible to think of the world as a place in which energy once was not an operant concept in physics. And yet, the physical concept as we know it only enters the realm of physics (and the English lexicon as the denomination of the present physical concept) in 1807, with the publication of Thomas Young's <u>A Course of Lectures on Natural Philosophy and the Mechanical Arts</u>. The emergence of the concept of energy in physics is the result neither of a "paradigm shift" in the narrowly scientific sense discussed by Kuhn, nor of a felicitous but anomalous thought experiment flying in the face of a powerful Newtonian hegemony. It is rather the result of a larger cultural change in the way that English thinkers viewed such issues as creation/causation, the mind-body relationship, the relative power of inductive and deductive logic, and the most efficacious approach to the problem of evil. This change ramified not only in science, but in literature and philosophy as well.

The present paper will attempt to situate the rise of energy as a chapter in a rising Protestantism's attempt to resolve all of the matters listed above as special cases of the problem of origins. The rise marks a shift from one of two antithetical accounts of origins in the Bible to the other, the second phase of an attempt to apply the Ramean law of truth to the resolution of paradox and aporia. Energy as a concept reverses Newtonian action at a distance rather than "correcting" the previous concept. Ultimately, when the rise of energy was found as wanting as action at a distance both were reincorporated in the decentered paradox that allowed for an inside and an outside, a particle theory of light and a wave theory, subject only to uncertainty and not to God's immanence for there to be a condition of comprehensibility. In the moment of energy's rise, however, the same sort of faith in the power to unlock the secrets of creation that characterized the earlier Newtonianism and its adherents characterized the party of energy.

Paul Theerman

Assistant Editor, Joseph Henry Papers, Smithsonian Institution

NEWTON AND THE IDEA OF "PURE SCIENCE": THE SIGNIFICANCE OF BIOGRAPH

In the early nineteenth century, scientists used biographies of past great men to buttress their claims to increased prestige and social status. But different men often wrote in support of different aims. Sir David Brewster (1781-1868) and Augustus De Morgan (1806-1871) both wrote biographies of Isaac Newton. But they came up with vastly different portrayals. Brewster's Newton was "modest, candid, affable, and without any of the eccentricities." Brewster was eager to promote the scientist as a gentleman. For him, Newton displyed the attributes of Romantic genius--a supernatural insight into the workings of the universe--coupled with the virtues of honesty, geniality, and piety.

De Morgan, while acknowledging Newton's achievements, preferred to talk of his sagacity, not his genius, deflating the idea that Newton was beyond the realm of mere mortals. More to the point, De Morgan pointed out Newton's contentious relations with Hooke, Flamsteed, and especially Leibnitz. De Morgan characterized Newton as having "a morbid fear of opposition." Further, De Morgan forced Brewster to acknowledge Newton's anti-Trinitarian views, a serious matter yet in Victorian Britain. De Morgan was concerned with breaking the link of genius and virtue that Brewster insisted upon.

The difference between Brewster's and De Morgan's views of Newton is in part due to their own life stories. Brewster was the public champion of the scientist; De Morgan relatively more concerned with private morals. But in addition, they had different ideas of the proper public position of the scientist, and their biographies of Newton reflected that.

Mark L. Greenberg

Department of Humanities, Drexel University, Philadelphia

BLAKE, TECHNOLOGY, AND LITERARY FORM

The nature, structure, arrangement, production, distribution, illustration, typographic conventions, political ramifications, modes of engagement, tactility, and psychological effects of books fascinated William Blake (1757-1827) throughout his life. Many of his works overtly concern books and their impedimenta, print technology, and ways of perceiving words and images communicated in print. In the <u>Marriage</u> <u>of</u> <u>Heaven</u> <u>and</u> <u>Hell</u> (1794) especially, Blake combines a self-parody of typographic form, an attack upon the work's own status as printed book, with sharp commentary on the commercial and technological forces conducing to produce printed works.

The <u>Marriage</u> is a book about books and reading: it specifies, satirizes, and ultimately obliterates a range of interacting texts, literary modes, and the ideas embodied in them. In the <u>Marriage</u>, moreover, Blake anatomizes minutely the very process of printing, exposing thereby the particulars of print technology and suggesting the ways it mediates ideas, images, and reader. Furthermore, the <u>Marriage</u> offers this anatomy in a variety of dynamically-shifting voices emerging from an alternative technology--composite art-- designed simultaneously to efface the limiting effects of typography while reorienting the ways we perceive. Text and design printed from etched copper plates and then hand colored allow Blake to fashion technology to his purposes rather than the other way around; and this "living form" as he calls it records the movements of his hand in time and space.

Blake's method of reproduction involves a process at once physically and psychologically apocalyptic: literally burning into the copper surface, biting into a body of ideas normally received in typographic form, and consequently transforming perceivers' minds. Blake's art thus enacts his fundamental belief that we become what we behold and that what we behold is determined by how we behold it. The medium--the kind of mechanical reproduction--is the message, linking perception with mental change. My approach to Blake's particular treatment of print technology and reading from within these complementary processes in the <u>Marriage</u> opens to more general observations about the nature of artistic reproduction and reception.

James Paradis

Massachusetts Institute of Technology, Cambridge MA 02139

On Science, Ideology, and Victorian Popularization

In Victorian popular science, we find a textual tradition that was counter-Humanistic in several important ways, which are examined in this paper. Rooted in the literary tradition of physico-theology, by which early naturalist-theologians secured the support of institutional religion, popular science became increasingly demotic in the nineteenth century, as vested intellectual interests fought for secular support by creating new readerships in Victorian England. On the one hand, writers such as Lyell, Faraday, Wallace, Huxley, and Lubbock were concerned in their popular works to link the sciences with traditional British culture; on the other, their popular science espoused new unities that were subversive. The underlying aim of this discussion is to examine some of the ingenious literary mechanisms and ideological arguments by which Victorians fitted science to the traditional Humanistic frameworks of understanding and to consider what this tells us about science, literature, and the Victorian imagination.

ALVIN C. KIBEL

HEAD, DEPARTMENT OF LITERATURE, MASSACHUSETTS INSTITUTE OF TECHNOLOGY

"DARWIN AMONG THE MACHINES"

Samuel Butler's Erewhon (1872) contains a long passage offered as a resume of the doctrines on the basis of which the inhabitants of his imaginary commonwealth long ago destroyed all machinery of any complexity and continued to prohibit its introduction and use. The passage is a tongue-in-cheek version of Darwinian theory, according to which machines have been utilizing human beings as extensions of their own organs of ingestion and reproduction and have been evolving at such a rate in consequence that they threaten to supplant mankind as a dominant species. Erewhon belongs to a kind of literary satire in which an author's own position with respect to a paradoxical play of ideas is hard to identify; moreover, Butler's later career as speculator in evolutionary notions landed him in the camp of the neo-Lamarckians. Hence treatment of this passage by literary scholars has tended to downplay the relation of its paradoxes to the substance of Darwinian thinking: Butler, it is said, did not properly understand Darwin and was, perhaps, only making a metaphorical appropriation of Darwinian themes in order to protest the growing predominance of machinery and mechanical procedures among the conditions governing everyday life during his time. In contrast, I suggest that Butler caught the speculative intention of Darwin's Origin more appropriately than most of his contemporaries and that the paradoxes of Erewhon are alive and well today, particularly in three fields of controversy: the so-called "units of selection" dispute in population biology, the place of functional or teleological judgments in science, and quarrels over whether digital computing machines can think.

Lance Schachterle

Professor of English, Worcester Polytechnic Institute

"WHAT REALLY DISTINGUISHES THE 'TWO CULTURES'?"

Nothing about C. P. Snow's essay has been more influential than its title. The phrase "Two Cultures" points to divergent patterns of education and socialization which irreconcilably divide scientists and humanists; the division has been extended further to propose intrinsic differences in how creative thought proceeds in the sciences and the arts. Yet in fact Snow's essay does not distinguish between the two cultures in terms of underlying intellectual activity. Rather Snow offers more superficial distinctions about attitude towards social change and progress. The evidence Snow brings to bolster his distinction suggests he is responding to his personal experience as a scientist and manager, instead of making larger claims for essential distinctions in mental activity.

In the more than quarter century since Snow's initial statement, thinkers from a variety of disciplines have offered sophisticated distinctions which point to similarities and differences between the "cultures" that Snow ignored. Scientists like Bohr and Heisenberg, and historians of science like Kuhn and Holton, have drawn attention to the crucial role of subjective judgment in science. Philosophers like Hesse and Rorty have delineated how the use of language shapes the kind and degrees of differences between fields and discourse. Further, recent fiction writers like Pynchon have negated Snow's criticism that "it is bizarre how very little of twentieth-century science has been assimilated into twentieth-century art."

John Woodcock

Associate Professor of English, Indiana University

SCIENCE AND SURVIVAL IN MARGARET ATWOOD'S LIFE BEFORE MAN

Images and themes associated with science play prominent and varied roles in Canadian writer Margaret Atwood's recent novel Life Before Man (1979). In an earlier novel, Surfacing (1972), Atwood set up a dichotomy between science, technology, logic, men, and America--which tended to be bad--and nature, feelings, earthiness, women, and Canada--which tended to be good. In Life Before Man that dichotomy is dissolved to some degree, but Atwood is still wary of science and the rational side of culture she associates with it.

Life Before Man is a dialogue between the forces of wholeness and life and dissolution and death in the struggle for emotional survival of three contemporary Canadians in their thirties. Each of the characters in the course of the novel gropes from an anesthetized state of death-in-life toward emotional openness and a vision of things that seems to make life minimally livable. Atwood in this novel is still very much interested in science, which she makes a major component of character, theme, setting, and point of view.

Lesje, the character who is a scientist, is a paleontologist at Toronto's Royal Ontario Museum. Her story embodies Atwood's more complex view of science. Lesje has chosen science as a haven from an emotionally destructive childhood and the social uneasiness that followed. Atwood presents Lesje's initial retreat into science sympathetically, but Lesje eventually grows unhappy with the shell she has built for herself, discovers the trace of a buried emotional life, and acts on it. She impetuously throws out her birth-control pills without telling her lover, hoping that this will produce the child he doesn't want but she feels she needs to survive. The technology of the pills and the cultural desire to control nature which they stand for have been blocks to her emotional integration.

The novel's major theme of emotional survival is repeatedly developed through threatening imagery of biological struggle, notably from the museum's natural history section. Science is also a part of the novel's setting in modern Toronto, with its museum and planetarium and sewage-disposal problems. Finally, the qualities of objectivity, abstraction, and reductionism--all of which Atwood associates with science--figure in the point of view of all three characters in a way that makes these qualities seem to be part of the culture the characters feel is hostile to survival. Atwood's science, in its emotional coolness and impersonality, in paleontology's fossil record and in astronomy's entropy and uncrossable distances, is on the side of death in this novel, as it was in Surfacing.

Donald R. Benson

Professor of English, Iowa State University

KANDINSKY'S RECONSTITUTION OF SPACE

If Wassily Kandinsky seemed less disturbed by the late 19th-century crisis over the nature of space than many of his contemporaries, still he managed, in both theory and practice, a reconstitution of space as radical as the Cubists' and perhaps Einstein's. The elements of this reconstitution were those his Symbolist tradition shared with late classical physics: atmosphere or ether, vibrations, and energy.

In physics the crisis had presented itself in questions about radiant energy, the structure of matter and its relation to energy, and the absolute measurement of motion. The crisis was resolved by the supposition of an ethereal medium capable of transmitting or expressing energy vibration or flow, of mediating, giving coherence to, even constituting the material and immaterial universe--a medium which, as Einstein later noted, brought space to life. In Impressionist and Symbolist art and theory the crucial problems were the realization of ambient light and color, and the relationships among the material world, consciousness and spirit. These problems were resolved by means of an atmospheric medium with the same essential properties as ether, which in the Symbolist version not only brought space to life but spiritualized it.

Atmosphere and its vibrations were important to Kandinsky's theory from the beginning as means of revealing and expressing the "inner sounds," the spiritual essences which were for him the true subjects of art, but he never fully resolved the contradictions between their given physical character and the spiritual character he ascribed to them. It was primarily through his discovery of the energies inherent in artistic elements and forms that he began a serious reconstitution of pictorial space, in theory and in practice. From the time of his early visits to the "magical," richly ornamented Russian peasant houses where he had learned "to live in the picture," he had pursued a painstaking discipline of thought and experiment to make color, weight, point and line extend pictorial space, draw the picture plane backward and forward, and allow the viewer to enter his pictures actively, enveloped by lines and shifting color patches, feeling the vibrations, hearing the inner sounds. Ultimately this "dematerialization" of the picture plane would generate an "immaterial" space to which the picture's inner energies and sounds were transferred and in which the viewer could participate. Here was a space, radically different from that of ether physics and Impressionist and Symbolist art, which would accommodate the phenomenal experience of an ultimately spiritual art.

Nelson Hilton

Department of English, University of Georgia

BLAKE AND THE PERCEPTION OF SCIENCE

Blake's apprentice work with the engraver James Basire (from 1772 to 1779) has often been discussed in terms of the "gothicizing" effect it is supposed to have had on his imagination. But Basire was also the official engraver for the Royal Academy for the Advancement of Science--his severely linear style suiting its needs as well as the gothic concerns of the London Society of Antiquaries-- and the case can be made as well for the imaginative impact of the more than one hundred illustrations to <u>Philosophical Transactions</u> which passed through Basire's shop during Blake's stay.

Blake's "An Island in the Moon" has been cited as early evidence of an anti-science outlook. But satire is not the same thing as rejection, and one may rather feel that Blake is making fun not of science but of those who have reduced experiments to parlor entertainment; the piece, in any event, turns on an informed awareness of medical, mathematical, and chemical concerns.

And Blake's reading of Swedenborg is usually put down to eccentric interest in yet another thinker far removed from the mainstream of empiricism. In fact, Swedenborg at the time of his "conversion" in 1745 had attained international stature as one of Europe's best-informed natural philosophers, and the science-based imagery Swedenborg uses throughout his theosophical work offers a rationale--as may be seen in his annotations--for Blake's interest.

Blake read intently and supplied illustrations for the poems of Erasmus Darwin; study of these and related works in the tradition of scientific didactic verse identifies some of Blake's apparent innovations as, rather, adaptions of that still neglected tradition. The difference is that in Blake the globules of man's blood or the globes of the heavens are not just personified--they also speak <u>in propria persona</u>.

During the last two decades of the century, Blake, as engraver and author, was closely associated with one of England's leading publishers of scientific and medical works, Joseph Johnson. Through Johnson Blake was in direct contact with a circle that included Joseph Priestley, John Bonnycastle, and Darwin.

Finally, Blake's overriding concern with the nature and operation of perception makes his poetic universe at least contiguous with the contemporary worlds of quantum mechanics and cognitive psychology. This association can be seen today in works of science fiction and science popularization.

Consideration of these issues can help us to understand more the place and play of "science" in Blake's work, and thus to appreciate more a concern and direction Blake announces in the closing words to that first modern epic, <u>The Four Zoas</u>: "sweet Science reigns."

Philip Appleman

Distinguished Professor of English, Indiana University

A Poetry Reading, from <u>Darwin's Ark</u>

The poems in <u>Darwin's Ark</u> (Indiana University Press, 1984) represent a lifelong commitment to the study of the works of Charles Darwin, and the profound satisfaction of knowing that one is truly and altogether a part of nature.

THE SKELETONS OF DREAMS

He found giants
in the earth: Mastodon,
Mylodon, thigh bones
like tree trunks, Megatherium, skulls
big as boulders — once,
in this savage country, treetops
trembled at their passing.
But their passing was silent as snails,
silent as rabbits; nothing at all recorded
the day when the last of them came
crashing through creepers and ferns,
shaking the earth a final time,
leaving behind them crickets,
monkeys, and mice.
For think: at last it is nothing
to be a giant — the dream
of an ending haunts tortoise and Toxodon,
troubles the sleep of the woodchuck
and the bear.

Back home in his English garden,
Darwin paused in his pacing,
writing it down in italics
in the book at the back of his mind,
 When a species has vanished
 from the face of the earth,
 the same form never reappears...
So after our millions of years
of inventing a thumb and a cortex,
and after the long pain
of writing our clumsy epic,
we know we are mortal as mammoths,
we know the last lines of our poem.
And somewhere in curving space
beyond our constellations,
nebulae burn in their universal law:
nothing out there ever knew
that on one sky-blue planet
we dreamed that terrible dream.
Blazing along through black nothing
to nowhere at all, Mastodons of heaven,
the stars do not need our small ruin.

CURT SIODMAK

Author of novels and motion pictures

SCIENCE FICTION, SCIENCE FANTASY AND THE WOLFMAN

As an author of Science Fiction I make a sharp definition between Science Fiction and Science Fantasy. Science fiction explores the possibilities of future scientific developments and discoveries and projects that condition on people of today. Science Fantasy speculates mostly in different social systems and social conditions, transported to other planets or imaginary space vehicles. The most celebrated proponent of Science-Fiction was Leonardo da Vinci, who anticipated the helicopter, the flying machines, the military tank and dozens of inventions which we now are taking for granted. For centuries the writing of Science-Fiction was looked down upon, though in my opinion pure Science-Fiction is every unsolved scientific idea. When the problem in solved, Science-Fiction becomes Science Fact.

The world of the imaginary dates back to Aristotle's POETICS, an exploration of Greek plays. He writes about HAMARTIA, an error in judgement resultion from a defect of the character of a tragic hero. Aristotle notes that his hero does not come into misfortune because of badness or rascality, but through some inequity or posetive fault. Most horror stories are founded on that basis, like the Frankenstein Monster or the Wolfman, who is destined to murder when the moon is full. Horror stories, Gothic tales, Science-Fiction and Science Fantasy are related.

Science-Fiction and Science-Fantasy are based on sharply defined themes. The question how to create a perfect social system are explored in Plato's REPUBLIC, More's UTOPIA, Huxley's BRAVE NEW WORLD, Skinner's WALDEN TWO. DONOVAN'S BRAIN tries to explore the limitation of the human brain, in HAUSER'S MEMORY, by the same author, the possibility of thought transfer is thematizised. Science-Fiction often explores the sociologies of different social systems (Orwell's 1984, Asimow's FOUNDATIONS etc.)

The pursuit of Science-Fiction and Science-Fantasy is built on questions which are strictly defined, like government systems on planets outside our galaxy. The search also opens up and tries to explain how ideas for writers originate, ideas which are generated by sociological factsors.

Science-Fiction and Science-Fantasy also prove the limitations of man's imagination, since his limited brain is incapable to interpret many secrets of nature. To explain to himself the seemingly inexplainable, Man created religions.

Imre Hronszky

A-prof. Inst. of Philosophy, TU/Bp, Hungary

Early and recent sociological approaches to the history of scientific cognition

A possibility to conceptualize society is to consider it as **practical totality**, in which the nature and history of all of its institutions is a **mediated** one. This viewpoint forbids an understanding of science as selfdetermining. From this viewpoint the 20th century science analysis seems the progressive destruction of the demarcationist presupposition system embodied in the justificatory view of science, commonly called the 'standard view'. Early in the 20th century the emergence of 'modern scientific **attitude' as such** became the explanandum on sociological grounds. But the scientific habit, having already emerged, was thought working under the 'logic of scientific cognition'/Veblen, Scheler/. This 'logic', it was presupposed, was derived from the social function of science and the lawlikeness of Nature together. Within that Marxist tradition which aimed to give full account of the practical nature of social being, Borkenau worked out an explanation of the emergence of 'modern science'.
Since the 70's this assumption of 'inner logic' has been gradually attacked from microsociological grounds. There have been efforts to show social elements lying behind and within the rational processes of scientific cognition, even working out scientific facts. At least as a demand a societal critical attitude to the 'microsociological' factors of scientific cognition has also been developed. Though the approach as whole is justified on the ground of the philosophy of science, nevertheless, in my estimation, there are numerous disturbing factors in the new approach. First of all, it has a strong appeal to a **conventionalist** understanding. Having sketched the development of the sociological analysis of science some alternative suggestions will be made, based on considerartions derived from the analysis of the double process of mastering and objectification of Nature.

Dr. P. G. Abir-Am

Lecturer, Tel Aviv University

"Re/consitituting meaning, social order and power in a transdisciplinary group: the Biotheoretical Gathering in England in the 1930s"

Transitory, especially transdisciplinary, groups are well suited, as a unit of historical analysis, to mediate between more traditional foci on either individuals or institutions. The Biotheoretical Gathering in England in the 1930s was such a group, composed of a dozen members who explored and sought to transform an informal, participatory and transdisciplinary discourse into a formal research institute associated with Cambridge University and funded by the Rockefeller Foundation. The non-conformist and marginal physicist, biologist, mathematician, and philosopher members (including among other J. Needham, J.D. Bernal, D. Wrinch, C.H. Waddington, J.H. Woodger, L.L. Whyte, M. Black, K.R. Popper) were impressed by the dual delegitimation of the classical scientific order following the quantum's and relativity's demise of the hegemony of classical physics in the 1920s. Their collective quest for new concepts, social orderings and power relations, reconstructed from meeting notes, correspondence and oral history, is examined for processes of interaction within the group, as well as objectified outcomes.

The activities of this group in constructing, validating and infesting with authority a new, transdisciplinary discourse on "mathematico-physico-chemical morphology", a known and catchier "molecular biology", are further interpreted in light of a new model of scientific action and order. This model aims to transcend the limitations (and accomplishments) of prevailing uni-dimensional models of scientific change, either "internalist" or "externalist", both "normative", "cognitive", and "authoritative".

One key feature of the new model pertains to its expanded range of explanation to include meaning, social order and power (hence the model's name as the 'Me-so-po-tamian Triangle') as simultaneous and mutually dependent, yet discrete and equally basic, variables. Another key feature pertains to the model's way of balancing process and outcomes by postulating new, 'meta-stable' states mediating between the micro-level of individuals and the macro-level of institutions, for each of the three, basic properties of interaction (i.e., meaning, social order and power). The predictive and interpretive qualities of the model are demonstrated while applying it systematically to the historical formation and dissolution of this group.

HERBERT MEHRTENS

Technische Universität Berlin

MATHEMATICS IN NAZI-GERMANY — PROFESSIONAL POLICIES IN DEFENSE OF AUTONOMY AND IN SEARCH OF LEGITIMACY

Reactions of mathematicians and their professional organizations to political pressures are analyzed. Autonomy and productivity of the discipline were threatened by Nazi-ideologues within mathematics and physics. Disciplinary organizations were - in collaboration with state-powers - able to marginalize such groups, which kept a residual role as ideological show-case of the discipline. This process helped establish the position of the discipline within the social system of Nazi-Germany. Social legitimacy and institutional status of mathematics were threatened by the purges, the duality of the power-system (state - party), and the anti-modern ideologies and policies in education. The institutions of the discipline reacted in various, functionally differentiated, ways. They offered, decreasingly in time, ideological service via school-mathematics and, increasingly, technical service through development of applied mathematics and education of technical specialists. International relations of mathematics were intensely but unsuccessfully defended. With this as well as with the loss of internal social coherence and ensuing power-struggles the discipline had to pay the price for survival and integration in the system of Nazi-Germany.

Peter Weingart, University of Bielefeld, West Germany
Harald Kranz, University of Bielefeld, Bielefeld,
West Germany

EUGENICS UNDER THE NAZI REGIME

It is part of the common perception of the role of science under the Nazi regime in general and of eugenics in particular that science was misused. The case of eugenics is instructive of a more complex relationship between science and politics. In the Weimar years the eugenics movement had undergone a split between moderate eugenicists and more radical racehygienists, which was decided in favor of the former. When the Nazis came to power this development was reversed and the racehygienists were reinstated. The new state implemented many of the measures demanded by the eugenics movement since the end of WWI, however, such as sterilization, health certificates as a precondition for marriage permits and financial support for large families. The majority of racehygienists failed to oppose the authoritarian aspect of these measures. -- The institutionalization of racehygiene as a scientific discipline, its growth and its role in policy-making, though impressive on the surface reveals the incapability on the part of the political administration to build up scientific competence in the field even on their own premises. Beginning in 1938 with a general trend toward a less ideological and more pragmatic orientation of Nazi politics race hygiene starts to decline. Much less than being misused by the Nazis, eugenics as a profession took an opportunist stand toward the regime and thereby entered a relationship of mutual exploitation with the fascist state.

Steven Shapin

Science Studies Unit, University of Edinburgh, Edinburgh, Scotland

THE LABORATORY AS PUBLIC SPACE IN 17th-CENTURY ENGLAND

We know remarkably little about <u>where</u> experimental science was done in 17th-century England. The physical and social nature of experimental spaces is, however, very important to our understanding of how scientific knowledge was actually produced. In 17th-century England it was widely held that experimental performances could yield matters of fact insofar as they were available to a witnessing public. By contrast, alchemical work was condemned because it was done in private places, with the consequence that its claims could not be validated. Thus, the study of the spaces in which scientific knowledge was produced is a topic for the historical sociology of knowledge.

In this paper I explore the nature of the emergent 'laboratory' in 17th-century England. I assemble some information about the places in which pneumatic experimentation was done, and I compare certain ideal pictures of the setting of experimental philosophy with verbal and pictorial accounts of its real circumstances. In particular, I focus on the extent to which experimental places were indeed public. Who was present? Who in principle could have obtained access? And who was excluded from experimental displays?

SIMON SCHAFFER

LECTURER IN HISTORY & PHILOSOPHY OF SCIENCE, CAMBRIDGE, U.K.

BOYLE AND THE GERMAN VACUUM

In *New Experiments Physico-mechanical* (1660) Boyle gave a detailed account of the construction of an air pump and of a set of experiments which, he claimed, led to matters of fact about the weight and spring of the air. The state of these matters of fact was intimately linked with the excellence of the pump. Boyle's description was allegedly detailed enough to allow the replication of his engine, and hence, of his trials within it. Any critic, Boyle reckoned, must try such replication before attacking the matters of fact produced in the engine.

Boyle was prompted to order such a machine by learning from Caspar Schott's *Mechanica hydraulico-pneumatica* (1657) of Guericke's experiments in pneumatics in Germany. Boyle pointed out defects in Guericke's device, and proffered his own machine as an obviously superior instrument in which the difficulties with Guericke's *antlia pneumatica* were decisively removed. Historians agree - the *machina Boyleana* has been seen as a clear progressive advance in instrumental technique, and an epochal emblem of the experimental philosophy in action.

But Boyle's claim to unquestioned superiority was a polemical one. In *Technica curiosa* (1654), Schott placed a full Latin translation of Boyle's book, accompanied with annotations by Guericke and by Schott on each trial. The Germans rejected Boyle's arguments for the virtue of his engine. Those points which Boyle said made his pump better were, for the Germans, marks of its obvious weakness. This applied, too, to Boyle's definition of a vacuum as a space from which air had been completely or almost completely removed. Schott said <u>all</u> the air could be taken from Guericke's engine. The issues at stake between Germans and English were those which linked the possibility of criticism in experimental work, the virtue of experimental instruments, and claims about the contents of nature.

In this paper I argue that in general claims about the competence of experimenters and their machines cannot simply be disentangled from debates about the contents of nature. Claims to matters of fact are simultaneously stipulations about the correct way of behaving in experimental work. The virtue of Boyle's instrument was a social accomplishment, and not imminent in his pump. Knowledge claims, therefore, involve the ordering of social skills.

Benjamin MATALON

Université de Paris VIII and Groupe d'Etudes et de Recherches sur la Science (EHESS), Paris, France.

Verification Practices in Physics and in Biology

The validity of the knowledge claims of science is said to be justified by the permanent criticism of the scientific community. But, if every result had to be verified by every scientist, the totality of the scientific activity would be spent in verifications. Nevertheless some authors have recently argued that, since the advent of "big science", there are less and less verifications, especially by replications. To replicate an experiment often takes too much time and money, and the pressures of competition makes this critical activity of very little interest. So, many results go unverified, and this leaves room for fraud or incompetence.

So it is interesting to know what are actually the practices of verification, and more specifically the practices of replication. Recent work in sociology of science have shown that replication is a problematic notion, and that the fact that an experiment is to count as a replication of another is open to discussion and negociation. Our aim is different : it is to understand what scientists do when they have decided to verify a certain result.

Our first finding is that, in physics at least, one seldom intends to replicate exactly an experiment. As physicists seem to trust the honesty and the competence of their colleagues, there is no point in doing exactly what they did. If something was wrong in the first experiment, it may be, for instance, because there was an artefact in the apparatus. So, one has to do another experiment, with less noise or with closer control. But a quite different experiment may aim at the same conclusion. It will be the generally accepted theory, if there is one, which insures the equivalence of the results. Besides, one can use a result, without verifying it directly : if it was wrong, one cannot go very far.

In biology, experiments take less time, are less expensive, and the variability of living material added to a certain mistrust among researchers, make replication more easy and more frequent. Many biologists, before using a new technique or before beginning to work on a new problem, repeat the most important experiments in the field. Even if their intention is not to verify them, the outcome is in fact the same.

K. Knorr-Cetina

Professor Fac. of Sociology, Univ. of Bielefeld

Experimental Work:
Comments from the Study of Contemporary Laboratory Sciences

The aim of this paper is twofold: First, I shall review relevant results of a series of recent studies (including my own) which, for the first time in the sociology of science, have chosen to study scientists at work through close, unmediated, systematic, ethnographic observation. Second, I shall draw upon my own present laboratory study to present some results on the topic of "how scientists think", on the role of apparatus and ma chinery in laboratory work, on the transformations and reconstructions inherent in the making of a knowledge claim, and on the structures of argumentation through which technical decisions are related to non-technicl considerations and constraints. These data will be derived from a comparative study of basic research in atomic physics and molecular biology which uses audio-recording, video-recording and interview material in addition to ethnographic observation. I shall attempt to elaborate and illustrate the program of constructivism in science studies, that is answer to the question in what sense knowledge appears to be constructed in and through laboratory work.

S. M. Razaullah Ansari

Professor, Department of History of Medicine & Science, IHMMR, New Delhi

THE DEVELOPMENT OF ASTRONOMICAL INSTRUMENTS IN INDIA

Though practical/observational astronomy in India could be traced back to the Vedic period (about a few millennia B.C.), yet the astronomy proper is considered to have started in the first few centuries A.D., when the five Sanskrit treatises Pañcasiddhāntika were compiled. In almost every siddhānta and its later commentaries there is a chapter on instruments, the Yantrādhāya. In Medieval India one finds about a dozen Sanskrit treatises exclusively devoted to instruments. Then in the Mughal period, the interaction of West-Central Asian Muslim and ancient Hindu scholarship led to the further development of astronomical instrumentation, when the Arabic-Persian sources on instruments, ālāt al-rasad, were available in India. The synthesis of the two schools of astronomy at the hands of Maharaja SWAI JAI SINGH II (1686-1743) crystalised in the shape of five observatories in a few Indian towns. There masonary and metal instruments, also 18th century telescopes, were installed. Only in the 19th century modern Western astronomy along with its excellent telescopes and mechanical clocks was introduced in observatories established at Madras, Bombay, Calcutta, Lucknow and Poona.

In this paper we attempt to trace briefly this growth of instrumentation from the earliest to modern times.

Ludolf von Mackensen

Prof.Dr., Hess. Landesmuseum u. Gesamthochschule Kassel

DEUTSCHER INSTRUMENTENBAU IM 16. JAHRHUNDERT
GERMAN INSTRUMENT-MAKING IN THE 16 th CENTURY

Verfeinerte quantifizierbare Wahrnehmungen haben in der Renaissance eine wichtige Rolle bei der Entstehung der neuzeitlichen Naturwissenschaft gespielt, insbesondere auf dem Gebiet der Astronomie.
Der Vortrag behandelt typische wissenschaftliche Instrumente für astronomische und geodätische Messungen im 16. Jahrhundert sowie ihre Vervollkommnung und Verbreitung in bedeutenden westdeutschen Zentren wie Augsburg, Nürnberg und Kassel.
Dabei werden auch Messergebnisse aus dem praktischen Gebrauch von Winkelmessinstrumenten und aus der Erprobung originalgetreuer Rekonstuktionen dikutiert.
Neuartige analoge Rechenhilfsmittel, wie der im 16. Jahrhundert neu auftauchende und von Jost Bürgi (1552-1632) verbesserte universale Reduktionszirkel (reduction compas), leiten über zu dem Bedürfnis nach digitalen Rechengeräten, wie sie dann erst im 17. Jahrhundert durch W. Schickard, B. Pascal und G.W. Leibniz erfunden wurden.

Paolo Brenni

Istituto e Museo di Storia della Scienza, Firenze, Italy

ITALIAN INSTRUMENT-MAKING IN THE 19th CENTURY

Italy is a country rich of XIX century collections of scientific instruments. Many of these collections are, still today, almost unknown; anyway there is an increasing interest for them. Many universities, institutes and schools (in Florence, Bologna, Pavia, Padua,etc) recently promoted restaurations and catalogations works of antique scientific instruments. Many of these instruments came from foreign workshop and firm (English,French and German) which constituted a massive presence in Italy during the XIX century. Anyway we can find instruments made in Italy by skilful craftsmen who were not able or didn't care about handing down their kneck. During the first half of the century we have,in fact, several constructors randomly dispersed and isolated,with few resources and often without inventiveness or particular talents.Their instruments, compared with similar instruments of foreign origin, appear to have been made at least some decades earlier. Sometime they were able and ingenious mechanics but without a solid scientific background.The scientist-constructors (äs G.B. Amici) were an exception in this panorama.
During the second half of the XIX century thanks to the changed social economical and political situation in Italy we witness the rise and growth on a national scale of a genuine industry of precision instruments brought about by efforts of enterprising and skilful constructors. However Italy was constantly very dependent upon foreign countries in this sector and never filled the technological gap.
In many cases the Italian instruments of the first half of the XIXth century are rudimental compared with the English or the French ones. Very often also the skilful constructurs made beautiful but unpractical instruments. In the second half of the century the firms producing instruments reached a good level of precision but (with the exception of the makers of geodetic instruments) they copyed without originality the catalogues and the products from abroad. It is today very difficult to find original documents, iconographical evidence or catalogues which furnish precise and detailed informations on the Italian makers and their instruments.It is therefore premature to attempt a detailed analysis of this scientific heritage, but we can give today a first list of Italian intrument makers of the XIXth century including all the informations now available about their workshops and their products.

Angelo Baracca

Università di Firenze, Florence, Italy

A DIFFERENTIATION BETWEEN "BIG SCIENCE v. LITTLE SCIENCE":
LAWRENCE AND TUVE, THE FIRST EXPERIMENTS WITH DEUTONS

At the beginning of the thirties a number of experimental groups undertook advanced research in nuclear physics in the U.S. They were composed of young physicists with new ideas. Some of them (Berkeley, Pasadena, Washington) developed and used particle accelerators. A careful analysis of their approaches and methods show deep differences between them and indicate the early roots of the differentiation between "big science" and "little" or "intermediate" science. The different experimental techniques and methods and theoretical premises between Lawrence's group at Berkeley and Tuve's group at the Carnegie Institution of Washington are considered in detail. Lawrence and Tuve were born in the same town and maintained a friendly relationship all along their life, but had deep scientific divergencies and bitter controversies. Lawrence created a very dynamic research team, aimed to the construction of bigger and bigger machines. He was personally a manager, able to rise funds from many Institutions, but had a very poor standard of scientific rigour, did not develope accurate experimental techniques, maintained a loose theoretical background and published hypotheses and results without serious checks. Lawrence's 1933 deuteron disintegration hypothesis was severely criticized and the majority of his experimental results published in those years turned out to be wrong. Tuve's group, on the contrary, always operated within the programs and the budget of the Carnegie Institution, accurately designed the accelerators for nuclear research, devoted a great attention to the experimental techniques and the rigorous check of the results. Tuve's warnings repeatedly, but unsuccesfully, reached Lawrence and he bitterly regretted the heedlessness with which the latter published his results: tihs is evident for the deuteron experiments. Tuve's group reached some of the most important results in nuclear physics. Tuve's dislike for the growing dimensions of research emerged clearly after the war, when he came back to direct the D.T.M. and abandoned nuclear physics because it "had turned into a business".

Willem Dirk Hackmann

Assistant Curator, Museum of the History of Science, University of Oxford.

SONAR RESEARCH AND SUBMARINE WARFARE: THE DEVELOPMENT OF INSTRUMENTS IN A MILITARY CONTEXT 1900-1940

Sonar research began as a desperate response to the U-boat menace in the First World War. Radical methods had to be devised, both organizationally and technically, to combat this new twentieth-century threat. Underwater acoustic detection seemed the most promising, as electromagnetic waves attenuate too rapidly in the sea. Hydrophones based on contemporary telephone technology were extensively deployed during the war, but more important was the work begun in France in 1915 on acoustic echo-ranging, in which sound waves were reflected off the submarine. In England the technique became known as 'ASDICS' - the acronym 'sonar' was not coined until 1942 at the Harvard Underwater Sound Laboratory. It became operational in 1919. In the inter-war years the Americans and the British continued to concentrate on these active sonar 'searchlight' techniques, while the Germans developed sophisticated passive listening systems, not for hunting submarines but for protecting their capital ships against torpedo attacks.

The distinction between pure and applied science is increasingly difficult to make, especially in the context of military research. Sonar development was governed by scientific, technological, bureaucratic, and political factors. In turn, its existence influenced governmental attitudes towards submarines. Germany focused on complex hydrophone arrays for her surface navy because she was forbidden to re-arm with submarines until 1935. By this time Britain had given up fighting for the abolition of this craft, partly because of an inflated confidence in her ASDICS anti-submarine measures. The US Navy took sonar development less seriously, because they did not have the British preoccupation with defending sea lines of communication against submarines. Thus, at the start of the Second World War, although the Americans had developed sophisticated sonar technology, only the British possessed an integrated sonar weapon system, passed on to the US Navy in Reverse Lend-Lease. This demonstrates how the direction taken by military scientific research, and the devices like sonar that were developed, were responses to national operational requirements.

World War I was the first time when large numbers of civilian scientists became involved with military research, setting the future pattern. Civilian scientists have played an increasingly important role in the technical affairs of the modern navy (as in all other military areas). One consequence has been the accelerated introduction of new devices into warfare. This development can be seen as a series of thrusts and counter-thrusts between ever more powerful weapons and their antidotes.

Yakov M. Rabkin

Professeur Agrégé, Université de Montréal

THE IMPACT OF SCIENTIFIC INSTRUMENTS ON RESEARCH

This paper is about the role of scientific instruments in the formulation, not only in the attainment, of research objectives. Based on the case study of Infra-Red Spectroscopy this paper integrates data from apposite secondary literature into a discussion of the changing functions of scientific instruments in the history of science. Originally considered mere means used in the pursuit of knowledge, scientific instruments have occasionally been shown to play crucial roles in removing cognitive bottlenecks. This paper attempts to broaden this picture to include the impact of the availability and of the cost of scientific instruments on the ways scientific problems are conceptualized and research programs are defined.

The transition of a scientific phenomenon from the role of an object of investigation to that of the scientific basis of a research tool is also discussed. This transition reflects the interplay of strategic and economic considerations, common in the generation of new technologies, and exemplifies an interesting instance of the science-technology relationship. New light is thus shed on the symbiotic reciprocal nature of this relationship which shows how a scientific idea can have a mediated effect on the advancement of science when embodied in a piece of new technology.

The mass production of scientific instruments represents a new and less conventional instance of the introduction of economic and technological influences into the orientation of basic science. While it amplifies the researcher's ability to manipulate nature, it also sharpens the experimental researcher's limitations in the choice of research topics. The input of the researcher into the manufacturing of the scientific instruments is also reduced with the advent of the mass production. Firstly, the instruments are serial products and their design can respond to collective rather than individual demands. Secondly, the heightened complexity of the instrument makes it exceedingly difficult for the scientist to modify the instrument to suit the needs of a particular experiment.

The industrialization of the production of scientific instruments also exercises an impact on the standards of legitimacy in the definition of research topics. The development of the instruments is therefore closely followed by the evolution not only of the nature but also of the scope of research. This impact has profound implications for the sociology of the scientific profession in the experimental sciences.

M.T. Wright, Science Museum, London, U.K.

J.V. Field, Science Museum, London, U.K.

MORE GEARS FROM THE GREEKS

The Science Museum, London, U.K., has recently acquired four fragments of a Byzantine portable sundial with associated calendrical gearing, datable to c.AD 500 (see our 'Gears from the Byzantines', *Annals of Science*, March 1985). The emergence of this instrument encourages speculation as to a possible link between the Antikythera mechanism (1st century BC) and the calendrical gearing described by al-Bīrūnī (c. AD 1000) (for the latter, see D.R. Hill, 'Al-Bīrūnī's mechanical calendar', *Annals of Science*, June 1985). Some of the features of the Byzantine instrument suggest ways in which one might reassess the Antikythera machine. For example, the making of wheels with "difficult" numbers of teeth must now be seen as practicable. On the other hand, the striking parallels between the Byzantine gearing and the mechanical calendar described by al-Biruni provide strong evidence for the transmission of this kind of technology from a continuing Hellenistic tradition into an emerging Islamic one.

PROFESSOR P E SPARGO

DIRECTOR, SCIENCE EDUCATION UNIT, UNIVERSITY OF CAPE TOWN.

The Role of Burning Glasses in Scientific Research - an historical overview for the period 1400 - 1900.

Most accounts of the history of burning glasses concentrate upon describing the spectacular results achieved by the use of burning glasses in melting metals and other refractory materials - or in allegedly destroying fleets of enemy warships ! However, such accounts present a somewhat one-sided view, for burning glasses have in fact a long and honourable history in serious scientific research.

Scientists - and in particular chemists - have always required sources of high temperature, and many areas of research have been dependent upon the development of improved high-temperature sources, such as furnaces, etc. However, for a very long period the highest temperatures available to scientists were not those produced by furnaces, but by concentrating the rays of the sun using lenses or mirrors.

Such devices had advantages other than simply the production of high temperatures, for they produced heat which was 'clean', in contrast to the interior of a furnace which all too frequently was contaminated by the gases and solid particles arising as products of combustion. The existence of a severely reducing atmosphere was thereby avoided. At the same time burning glasses had the almost unique ability for their time of being able to produce areas of high temperature <u>inside</u> a transparent glass containing vessel - either at atmospheric pressure or under vacuum conditions.

In spite of these advantages burning glasses also suffered from a number of serious disadvantages which severely limited their wide-spread use in research. Firstly, they were extremely weather-dependent, which was no mean problem in many of the cloudy (and smoky) areas of Northern Europe. Secondly, their energy output varied quite sharply according to the season of the year. Thirdly, although they were capable of producing high temperatures, their actual <u>energy</u> output was quite small when compared to a large laboratory furnace, which meant that only relatively small volumes of material could be raised to high temperatures.

Ferenc Szabadváry
Director,
Hungarian Museum for Science and Technology, Budapest

HISTORICAL DEVELOPMENT OF CHEMICAL LABORATORY
AND ITS EQUIPMENT

Origin of the chemical laboratory. The historical scenes of chemical research work. The way of chemical analysis from qualitative detections till the beginning of quantitative determinations. The first analytical devices for quantitative purposes. The penetration of optics and electricity in the chemical laboratory work in the XIX. century. Optical and electrical analytical instruments. The instrumental development at the beginning of our century. The actual change of the usual view of an analytical chemical laboratory, domination of electronics.

Trevor H. Levere

Professor & Director, Inst. for the History & Philosophy of Science & Technology, University of Toronto

SCIENCE, EXPLORATION, & INSTRUMENTS IN THE CANADIAN ARCTIC

The end of the Napoleonic Wars was followed by the rapid expansion of Arctic exploration, first primarily directed at the discovery of a North-West passage, later increasingly concerned with national prestige and with the expansion of scientific knowledge. The Canadian Arctic was part of the Empire; the archipelago was not transferred to Canada until 1880. Scientific work there for most of the nineteenth century was thus imperial rather than colonial science, and was principally directed by the Admiralty in London, with occasional assistance from the Ordnance; Sabine, for example, came from the Royal Artillery, as did Henry Feilden, naturalist on the Nares expedition of 1875-76.

These expeditions were involved in numerous branches of science, including meteorology, oceanography, and magnetism, essential for navigation, but problematic in iron ships and high latitudes. Investigation in these sciences depended upon the development of appropriate instruments -- Ross's deep sea clamm, Six's registering thermometer, and Fox's dip circle -- all of which were tested extensively in extreme conditions during the Arctic voyage.

This paper gives especial consideration to the instruments used on the expeditions of J.C. Ross, W.E. Parry, J. Franklin, and G.S. Nares, and their role in determining the scientific significance of those expeditions. Ross was keenly interested in many facets of geophysics and marine science. Parry spent almost more time preparing himself for the scientific aspects of his first command than for the more narrowly naval ones, frequently sharing in magnetic and other observations. Franklin was also much involved with the scientific side of his voyages; it was his advocacy of Fox's instrument that led, through Beaufort, to its adoption by the navy, to its use on Back's expedition, and thus to its major role in nineteenth-century naval science. Nares was in command of the Challenger in the Southern oceans when he learned of his transfer to H.M.S. Alert on Arctic service; he took with him experience in oceanography, including contributions to the design of dredging equipment. The interaction between the Navy, the Royal Society, and the instrument makers who supplied them, underlies a century of science in Arctic exploration.

ALAN STIMSON

CURATOR, NAVIGATION, NATIONAL MARITIME MUSEUM, GREENWICH, LONDON SE10

NEW EVIDENCE FROM WRECKS

Museums have traditionally gathered their collections from the salerooms and from the attics of society. The basis of most collections was very often 'fine art' orientated and not always representative of a true cross-section of the market at any one time.

Much of the detail history of nautical science instrumentation is founded upon these collections and documentary evidence, for the most part, composed by the élite of the profession. Since World War II an increasing amount of fresh evidence has been available to the historian recovered from numerous wrecks scattered around the world. Opportunities are now available to confirm or contest earlier suppositions by recording and analysing the many artefacts recovered from wrecks.

Instrument finds from BATAVIA in the early 60's to MARY ROSE at the present time have contributed new evidence. Some of the more important discoveries are discussed in this paper.

R G W Anderson
Director, Royal Scottish Museum, Edinburgh, Scotland

THE ARCHAEOLOGY OF CHEMICAL PRACTICES

Historians of Science have not, generally, capitalised on the highly developed techniques of archaeologists and art historians in examining artefacts to provide evidence for historical studies. An area where such approaches might be useful is in the investigation of early chemical processes. In particular, the development of distillation methods, which require specially developed apparatus, is susceptible to elucidation by these means.

A few early pieces of evidence come from prehistoric societies (where there is no alternative to examining the artefacts). A number of somewhat later Near Eastern alembics and cucurbits are to be found in museums, though their provenances are mostly unknown. The main published evidence of recent years has been provided in reports by medieval archaeologists who have discovered material at castles, monasteries, glasshouses and kilns at various European sites.

This paper surveys the evidence and shows how it differs from accounts of the evolution of distillation which are based purely on documentary sources.

Gerard L'E. Turner, Museum of the History of Science, University of Oxford, Oxford, England

Historical Evidence from the Decoration on Scientific Instruments

Archaeologists and historians of the decorative arts well understand the value of the decoration on artefacts as evidence towards establishing date and place. Such techniques are only now beginning to be employed in the critical study of the structure of the instrument-making trade, and in positive dating. Such studies also relate scientific artefacts to the cultural context of which they were the product, linking them to styles in typography and printing, and in furniture.

A detailed examination was made of the gilt decoration on the leather or vellum body tubes of nearly 100 microscopes and telescopes of the seventeenth and early eighteenth centuries. This decoration is of the same type as that found on book bindings of the period. The study revealed that, contrary to previous accounts, the shapes of the decorative stampings were not peculiar to any particular maker and that three stylistic periods could be distinguished, so providing an aid to dating. That instruments marked with different makers' names showed that there were specialist tube makers working for the retail trade.

A study of the letter forms in the engraving on a number of signed English Elizabethan instruments has made it possible to assign makers' names and dates to unsigned instruments. This research has revealed that there are twice as many Elizabethan instruments preserved as was realized only two years ago, giving a current total of over 80. The identification of such numbers and variety has led to a reassessment of the English instrument-making trade in the second half of the sixteenth century.

Dr. Penelope Gouk, St. Hilda's College, Oxford, England.

THE SIGNIFICANCE OF IVORY

That ivory was frequently used along with brass, silver and wood in the manufacture of scientific instruments in early modern Europe is a measure of the value placed on such objects. With the help of navigational instruments vast areas of the world were rapidly being opened up to the possibility of commercial exploitation. Astronomical and mathematical instruments enabled men to keep time and to measure and quantify the natural world, thereby imposing some degree of control over their environment. Apart from having a functional role to play, these instruments symbolised the power, wealth and status of their owners, and would often appear in the Kunstkammern of kings, aristocrats and the wealthier merchants.

Ivory had always been a precious commodity, used in the production of luxury goods. Imported from Africa and India via the new sea trade routes it was sent to major commercial centres wealthy enough to support the manufacture and distribution of such items. In the fifteenth century these had been mainly in Italy, but by the sixteenth century the economic centre of gravity had shifted to the South German cities of Nuremberg and Augsburg. These became major centres for the production of valuable technological goods such as weapons, clocks and scientific instruments, as well as artistic objects.

It is not surprising to find that ivory came to be used in the manufacture of instruments, since one of its advantages is that it is resilient, but soft enough to lend itself to turning, carving and engraving. Microscopes made all or partly of ivory were produced because of the ease of turning them on an ornamental lathe. They could be made by artistic turners as well as instrument makers. These men also produced anatomical figures and models to demonstrate the workings of organs such as the eye and ear. Other types of instruments made of ivory include quadrants, pillar dials and barrel compasses. The most popular type of ivory instrument, however, was the diptych, a form of portable sundial. The major centre for their production was Nuremberg between 1500 and 1700.

It is in the study of such instruments that the additional significance of ivory is revealed. Because of its durability as a material, a high percentage of these diptychs have survived, produced by a relatively small number of known makers. Their marks and designs can be studied in order to reveal techniques of production and guild practice as well as to establish precise connections between instrument making and the decorative arts. In ivory instruments, art, technology and science are combined to achieve a balance between aesthetic and functional demands.

Alain BRIEUX, Paris. Corr. Member A.I.H.S.

Georges IFRAH, Paris. Alain SEGONDS, Paris. CNRS.

EVIDENCE FROM CALCULATING DEVICES

The evidence of sophisticated instruments of calculation in the Romans' culture comes from a few samples known in the previous literature and remaining samples preserved in Museums.
Athanasius Kircher, Lorenzio Pignoria, and Marc Welser have written on these instruments and appear to have seen some of them.
Some of these instruments are now preserved in the following Museums or Libraries : The British Museum, the Bibliothèque Nationale of Paris, the Museo dei Thermi in Roma and one, recently discovered and identified by the authors, now belonging to the IBM Europe collection in Paris.
The evidence of the use of these instruments by the Romans reckoners is also attested by one stonegrave in the Museo Capitolano in Rome.
George Ifrah, the author of the book recently published in this country under the name "From one to zero"* discussed the history of numbering by Roman numerals and explained the evidence of datation of the IBM sample by knowing the evolution of that notation.
The authors will try to trace these samples in the literature and in the collections.

(Slides showing all these instruments and sources will be provided)

* Vicking Press N.Y.

D.E. ALLEN

Winchester, U.K.

THE STRUCTURE OF NATURAL HISTORY JOURNAL PUBLISHING IN MID-NINETEENTH CENTURY BRITAIN.

Scientific periodicals of a broader and less ponderous character began to appear in Britain in the 1820s. But advances in printing technology and, later, fiscal reforms soon lowered costs so much that a popular market became feasible. The high profitability of his <u>Gardener's Magazine</u> led J.C. Loudon to launch the <u>Magazine of Natural History</u> in 1828, the early success of which attracted over-many imitators. The market was smaller and more inelastic than generally supposed, and minor misjudgements quickly proved fatal. These included pricing too low, publishing too frequently, paying contributors, overpaying editors. Making the content less demanding was liable to lose more readers than it won; aiming high tended to be uneconomic, similarly. Commercial practicality and the needs of science seemed to be in permanent imbalance. The growing specialization in science made matters worse, for individual disciplines were for long too small to support a reputable journal unless there was a specialist society to subsidise it. It was not till the 1860s that botany and entomology each acquired commercial journals of a long-term viability.

William H. Brock

Reader in History of Science, University of Leicester

THE CHEMICAL NEWS 1859-1932

The *Chemical News* was founded by the 27-year old William Crookes on 10 December 1859 and was wound up in October 1932 after its 3781st issue. Although not the first English commercial journal devoted to the chemical sciences (it incorporated the *Chemical Gazette* of 1842-59, for example), what was distinctly new was its *weekly* schedule, this enabling it to act as a very rapid channel of chemical communication. As such it was a model for Norman Lockyer's *Nature*, first published in 1869.

Like Lockyer, Crookes was both a scientist of great originality and a journalist with sound commercial sense. By July 1860 Crookes was breaking even financially with 870 copies a week. In 1861 he raised the price from 3d to 4d (it only moved to 6d in 1920) and tried to inject more capital into the venture, becoming the sole owner in the process. By 1900 it was bringing him an annual income of £400 pa.

In 12 to 14 closely-packed double-columned pages, which were indexed every 6 months, Crookes and one assistant provided verbatim accounts of scientific meetings, abstracted articles from continental journals, offered a venue for airing disputes, issues and speculations, and publicised his own exciting work on thallium, the radiometer, cathode rays. Changes of tone may be noted: more elementary articles in the early years, great concern with education, public health and professionalization in the 1870s (an annual guide to schools of chemistry was published from 1863), greater concern with theoretical chemistry from the 1880s.

The journal never fully recovered from Crookes's death in 1919, though faithfully continued by his laboratory assistant, J.H. Gardiner, until 1923 and by the researching school chemistry teacher, J.G.F. Druce (1894-1950), until 1930. With serious competition from the weekly *Chemistry and Industry* from 1923, *Chemical News* ceased to carry primary communications and more and more assumed the air of a popular science magazine - a process hastened by the final editor, H.W. Blood-Ryan, who lost its solvency in a final fling of high quality printing and illustration of the vacuous.

Jean Guy DHOMBRES, Professor, Université de Nantes (Mathématiques).

THE EVOLUTION OF PUBLISHING IN THE EXACT SCIENCES, 1750-1850.

If the popularity of science is well established in France during the French Revolution and the Empire, and was much more than it was before or immediately afterwards, were all sciences equally popular, or equally practiced ? We intend to study the interplay between science, as it is constructed, and its popularization through books and textbooks. To do this, we try to present and to scrutinize all kinds of statistics which can be produced from archives about scientific books, and scientific papers being printed in specialized journals, comparing the number and the quality of editions, according to different subjects in the exact sciences. Looking on a rather long and rich period of time, from 1750 to 1850, and comparing between different European countries, we may draw some conclusions concerning the practical as well as the ideological rôle played by the sciences during this century.

A clear result of this quantitative study is to stress the particular part played by mathematics, especially during the golden period for scientific publications in France (from 1797 to 1819). The steady expansion of mathematical literature with an average of 14% of all published books in science over the period, cannot be explained on the basis of a richer activity in research. Other sciences show far lower ratios of published books to research papers and a lack of eagerness to popularize recent results in books. This fact being established, various factors may be provided to yield an explanation.

From such an apparently dull study of publishing in the exact sciences, some hypotheses can also be put forward about the evolution of the structures governing scientific communities, from the Encyclopedic period to the time of positive science, at the dawn of the industrial era. Particularly clear is the trend toward self-sufficiency within scientific communities, in the direction of cutting all links with other cultural domains. Typical is the evolution of textbooks, over a century, showing a far clearer distinction between scientific subjects, the progressive but noticeable neglect of applications from the domain to which a book is devoted to other domains, and adoption of few "classical manuals" which serve as patterns for most of the other textbooks. The architectural point of view, so characteristic of the mathematical sciences, is emphasized.

Kara-Murza, Serguei, Dr. of Science in Chemistry
USSR Academy of Sciences, Institute of the History of
Science and Technology, Moscow, USSR

Journal publications: an approach to the study of the evolution of the cognitive structure of a discipline

The cognitive structure of a scientific discipline or a research area is considered as a system of closel interconnected elements: scientific facts, theories an experimental methods. The structure of science is a co bination of fields with high-level cognitive consensus and "diffuse" research themes with low-level consolida tion. The cognitive structure of compact fields is represented by the clusters of key papers (H. Small, B. Griffith). The cognitive structure of a discipline is reflected in the body of journal references. If the subject orientation of the cited journals is known, th evolution of the cognitive structure may be traced fro the dynamics of the structure and the age of reference of the citing journals.

Experimental methods described in journal publications serve as an appropriate indicator of evolution and transformation of the cognitive structure. Both cognitive and social factors are reflected in the dynamics of the use of the main methods and changes in the structure of the whole methodological complex. The cases discussed is the evolution of the structure of the methods of the organic synthesis and the set of physical methods of organic chemistry and biochemistry after the second world war.

Christoph Meinel

Universität Hamburg, Institut für Geschichte der Naturwissenschaften, FRG

COMMUNICATION AND INTEGRATION - CHEMICAL JOURNALS AND THE INTERNATIONAL TRANSMISSION OF KNOWLEDGE IN NINETEENTH-CENTURY EUROPE

Scientific communication forms the basic structure of any scientific community. In chemistry, at an early stage, the specialized journal became the medium that brought about and maintained the scientific and social identity of the discipline. Despite major transformations the journal retained its key position throughout the century, and only after 1900 did the first signs of an information crisis appear in chemistry, leading to the still continuing debate about the future of the scientific journal.

In 1778 chemistry created a successful, discipline-oriented journalism that soon became a model for other disciplines. Pressed by economic conditions and the increasing amount of factual information, chemical journals soon went through a period of experimentation in an efford to fulfil their twin functions as both a repository and a platform of critical feedback. In this study I shall concentrate on the most vulnerable, and therefore the most delicate, aspect of scientific journalism: the exchange of information across national or linguistic frontiers. Journal editors tried several solutions to cope with this problem which were partly quite successful as case studies of the transmission of information reveal. Its efficiency can roughly be estimated from the proportion of articles published in more than one language. Here significant differences between Britain, France and Germany point to specific conditions within their chemical communities. However, during the first half of the 19th century, the proportion of papers in journals which were translated or published simultaneously in different languages was high. Social integration on a national scale and transnational communication complemented each other.

In the second half of the century, the foundation of national chemical societies reveals a change within European chemistry. Due to the centralizing and mediating functions of these societies, translated articles and multiple publication disappeared almost completely in the 1870's, and more standardized forms of journals with a strict separation between substantive and abstract function emerged. At the same time the entire communication pattern changed. The ideal of a republic of letters was replaced by nationally organized communities, linked to each other not through the individual scientist, but rather through the national organisations. Consequently, proposals to create in chemistry one single international abstract service, in order to avoid fragmentation into several national communication networks, did not succeed. Thus the multifaceted, personalized and mostly unstable exchange of information of the early 19th century was transformed into an institutionalized and stable, though sometimes inflexible and impersonal communication pattern dominated by the national chemical societies. These changes and developments within the European chemical community were clearly influenced, and are historically well documented, by the most important structural element of this scientific community: the chemical journal.

János Farkas

Professor of Sociology, Technical University, Budapest

Inconsistent Interpretations of Marxian methodology in the
History of Social Sciences

The marxist sociologists of science usually derive the social
essence of science from the working process. The methodological
problems of this argumentation begin when they depart not from some
given historical form of production, but from a structural description of work which they assume to be the model of social practice
as a whole. Consequently, they are compelled to link all the
important social activities (e.g. science, too) directly into this
notion of work, to derive them "genetically" right from the concept
and model of labor. Furthermore, they are constrained to mention
only as an example - or to ignore it altogether - the real social
history of the formation of social objectivations, activities and
spheres. In this manner are we lead to the history of origin of
the sciences.

Gad Freudenthal

Centre national de la recherche scientifique, Paris, France

Hélène Metzger's Hermeneutics of the History of Science

Hermeneutics is the discipline investigating the conditions of the possibility of interpreting texts. Following Luther, classical hermeneutics (Dilthey, Schleiermacher) maintained that the objective understanding of an alien thought in a text is possible. This view was vigorously challenged by H.-G. Gadamer (in the wake of M. Heidegger): interpretation, he holds, necessarily depends on and involves a fore-understanding, a projection of the interpreter's own horizon ("prejudice") unto the text. Therefore, "if we at all understand, we understand differently."

Hélène Metzger seems to have been the first to reflect on whether the historian's understanding of past scientific texts can be objective and thus to propound a hermeneutical theory of the history of science. Arguing from the underdetermination ("Duhem-Quine") thesis, she held 1. that any set of texts is always compatible with different interpretations, and 2. that, consequently, it will be each historian's partly a priori epistemology that will command the historical structure into which s/he will embed the texts. An epistemological "prejudice" thus underlies all history and, indeed, "the facts as interpreted by the positivist doctrine lend support to the positivist." Metzger, like Gadamer, rejects the Enlightenment view that the notion of knowledge implies objectivity, i.e. the the elimination of all preconceived ideas. Yet Metzger at the same time tries to avoid seeing the historian of science as "the soldier of a philosophical theory," to steer clear of both positivism and relativism. In this task her success (like that of present-day philosophers) was very limited.

Maurice A. FINOCCHIARO

University of Nevada--Las Vegas, U.S.A.

COMMENTS ON THE METHODOLOGY AND PHILOSOPHY OF HISTORY OF SCIENCE

My comments examine primarily the papers dealing with problems in the Congress Symposium on the methodology and philosophy of history of science. Secondarily, I discuss relevant background issues and ideas. The methodology of history of science may be conceived as the study of the proper methods for writing about and doing research in history of science. The philosophy of history of science may be taken to include the just mentioned methodology, as well as the examination of such topics as the nature of the developmental patterns, if any, in the evolution of the various scientific disciplines.

Joseph Agassi

Professor of Philosophy, Tel-Aviv and York University, Toronto.

TWENTY YEARS AFTER

Scholasticism is a system of patchworks created for the purpose of staying within one paradigm despite criticism yet without ignoring criticisms. Historians like to parade series of paradigms as they occurred in history, though philosophers may be uneasy about the arbitrariness of the choice of one. Rank-and-file intellectuals who simply accept paradigms from their leaders avoid this problem of choice, but how do leaders choose? They try to avoid scholasticism as much as they can.

My <u>Towards an Historiography of Science</u> of twenty years ago presents a few paradigms of how to present the history of science and proposes that the Popperian one is best among them. This proposal of mine was neither endorsed nor rejected by the field's leadership, since that field has no leadership. Yet the Popperian paradigm is now sufficiently popular to be one of the competitors in the field.

For my own part I think Popper's explanation of his rejection of scholasticism is not valid. I think science needs frameworks or paradigms, somewhat the way Emile Meyerson proposed, and a critical attitude, akin to what Popper proposed. Alexandre Koyré is still the best example of one whose work embodies both these ideas though not systematically enough. We want to try to integrate our pictures of past scientist, attempting - but never fully succeeding - to integrate their philosophy of nature, philosophy of science and science. A historian should openly record his disagreements with his heroes in all three of these areas.

Henryk Hollender, Uniwersytet Warszawski, Warsaw, Poland

Eugenjusz Olszewski, Politechnika Warszawska, Warsaw,
Poland

REGULARITIES IN THE EVOLUTION OF PARTICULAR SCIENCES

The course of changes within each branch of learning depends, among other things, upon that branche's susceptibility to external /social/ influences. If we arrange sciences accbrding to the degree of their susceptibility to such factors we get a continuum at whose opposite ends we find respectively mathematics with formal logic and historical sciences. Beetween them, in the order of their growing dependence cn external factors, are placed the sciences dealing with systems and information, natural, technological, agricultural, medical, and social ones.

The position of a particular science within the continuum is usually determined simultaneously by: 1. The degree of its subject's variability /starting with the subject being created through the process of mental investigation to the subjects being changed in historical time/; 2. The extent to which the notion of the past is included in a given science /from its absence to the situation when it is one of its essential elements/; 3. The degree of abstractness; 4. The degree to which its successive statements correspond to each other /from a complete correspondence to the one impossible to determine/.

Since this collection of features does not change within the continuum isomorphously, the latter is not a simple line-up. At some of its sections there may be parallel sciences or their ramifications.

An essential factor in the evolution of particular branches of science is a quasi-synchronical correspondence between them. This correspondence is determined mainly by the combination: questions and answers /borrowing and lending/ concerning the methods or the subject matter. All the branches transmit their content both ways, toward the "mathematical" and "historical" ends of the continuum, with the latter end not being distinctly separated from various groups of non-scientific learning /including ideology/. Knowledge can be also transmitted through a ramification of the continuum that passes round a group of sciences. Thus for instance the technological sciences are linked to practical knowledge without social sciences as the go-between. The continuum constitutes a model of scientific knowledge, being a system in which particular elements show alternately greater variability which they lend to the remaining ones through their interdisciplinary contacts. Occasionally the changes may be started by a concentration in a branch of science of many other branches' methods and problems.

Marcello Pera

Department of Philosophy, University of Pisa, Italy

A DIVORCE FOR LOVE BETWEEN THE HISTORY AND PHILOSOPHY OF SCIENCE

0. As a result of the "historicist" revolution, a debate has arisen as to whether and to what extent the history and philosophy of science are related."Intimate relationship or marriage of convenience?". A radical view will be put forward here,as far as methodology is concerned It will be argued that a separation between the two partners benefits both of them and makes room for an alternative view of science, different from "science as impersonal knowledge", on the one hand, and from "science as propaganda", on the other. The whole argument will be developed in three stages.

1. The main theses of the historicist view will be briefly presented and examined. In particular, attention will be focused on the historical authentication of the methodological standards or models or criteria of rationality and progress. It will be argued that this kind of justification raises difficulties, which are numerous and serious enough to discredit it and to recommend a different approach.These difficulties are: circularity, inconclusiveness, irrealizability,arbitrariness, normative inefficacy and historical adhocness.

2. A different sort of justification,called "axiological vindication", will be proposed. According to this view, methodological standards are introduced as hypothetical imperatives subordinated to a set of goals considered as primary cognitive values. The historicist argument will then be reversed: it is not because a method M agrees with a sample from the history of science S that M has to be accepted;on the contrary, because M is acceptable, S is scientific or rational. Independent grounds of acceptability for a set of goals and a hierarchy of these goals will be introduced and it will be maintained that, under this construal, methodology maintains a normative status without forfeiting its relationships with the history of science.

3. An examination of the logical link between goals and norms shows that any methodological code has an irremediable "open texture". While discouraging sharp, rigorous,clear-cut demarcation criteria and explications, this theorem suggests a new way of looking at scientific endeavour, namely the view of "science as argumentation". Certain advantages of this view will be examined. On the supposition that these really exist, the moral may be drawn that, in the field of the history and philosophy of science, divorce is better than marriage. The extrapolation of this conclusion to other fields is left to the reader.

Tögel, Christfried
Bulgarian Academy of Sciences, Sofia, Bulgaria

THE HISTORY OF PSYCHOLOGY FROM A STRUCTURALIST POINT OF VIEW

The structuralist view of scientific theories, introduced by Joseph Sneed (J. Sneed, The Logical Structure of Mathematical physics, Dordrecht 1971) and elaborated by Wolfgang Stegmüller (W. Stegmüller, The Structure and Dynamics of Theories, New York 1976) and Wolfgang Balzer (W. Balzer, Empirische Theorien: Modelle Strukturen, Beispiele, Braunschweig 1982) is applied to psychology. The paper shows that the structuralist interpretation of psychological theories leads to a new understanding of several aspects of the history of psychology. The fact that there is no need for scientific theories to be made immune to falsification, because they are immune, is illustrated by the psychoanalytic theory of personality as it was established by Sigmund Freud in 1923 (S. Freud, Das Ich und das Es, Leipzig, 1923). This example shows that the structuralist interpretation of Thomas Kuhn's concept of science can explain the ostensible irrational behavior of psychologists: adhering to theories (for example: to psychoanalytic theories), despite the falsification of their empirical hypotheses.

Beside this, from the structuralist point of view historical continuity in science is not connected with truth of theories, but (1) with a person p_1, who had previously posited the core of a theory, the set of paradigmatic examples and had been the first to undertake a successful expansion of the core of the theory, and (2) with a person (or group) p_2, ready to accept the set of paradigmatic examples.

In the case of the psychoanalytic theory of personality its core is the idea of the structural frame of personality, consisting of Id, Ego and Super-Ego. As paradigmatic example Freud considered the aetiology of neurosis, the first succsessful expansion of the core of theory was its application to schizophrenic psychosis and the psychoanalytic community (p_2) is determined by the acceptance of the paradigmatic examples, given by Freud.

Conclusion: The structuralist view of theories, elaborated in the main in the field of physical sciences, allows a new interpretation of the historical aspects of psychology too.

Manu Jääskeläinen, PhD

Chief of Educational Affairs, Finnish Medical Association

Development of Psychology in Scandinavia

The Scandinavian countries Denmark, Finland, Iceland, Norway, and Sweden are all countries with partly different and individual, partly common historical, cultural, and economic characteristics. Total population in Scandinavia is c. 22,6 mill. people. In this area, there are 50 universities or scientific institutes for higher education (with postgraduate and research resources). According to Edwin G. Boring, "one must beware of placing too much emphasis upon the relation of geography to philosophical point of view." (E.G. Boring, A History of Experimental Psychology, 1957, p. 440).

In studying the history of scientific psychology in Scandinavia, one must take into account the individual characteristics of these countries as well as the common trends caused e.g. by the scientific and cultural cooperation between them.

First institution in Scandinavia in this academic field was the Psychological Laboratory at the University of Copenhagen, founded by Alfred Lehmann (1858-1921) in 1886. He studied at Wundt's laboratory, and was strongly influenced by German psychology. He was an original scientist with multifarious interests, and was internationally well-known. His successor was Edgar Rubin (1886-1951), famous for his studies in the field of perception. In Sweden, first institute was founded in 1902 at the University of Uppsala by Sydney Alrutz (1868-1925). He was internationally known for his studies of the sensations of the skin. In Norway, the Institute of Psychology at the University of Oslo was founded in 1909 by Anathon Aall (1867-1943), a professor in philosophy. His work was continued by Harald Schjelderup (1895). Schjelderup's chair was converted into a chair in psychology in 1928. He was strongly influenced by psychoanalysis. Recently, cognitive psychology has attracted an increasing number of investigators. In Finland, first Institute of Psychology was founded at the University of Turku (Åbo) in 1922 by Eino Kaila (1890-1958) who is the "founding father" of academic psychology in this country even if earlier works e.g. in psychophysics and experimental psychology were counducted and published. The influence of German and Scandinavian, especially Danish, experimental psychology and Gestalt psychology has been remarkable.- In the paper, common and national trends will be further analyzed

dr. György Kiss

Research psychologist, University of Technology Budapest

Development of experimental psychology in Hungary

Experimental psychology has not started to develop from an academic, university environment as, for instance, in Germany, its basis was established by demands of practice.
Hungarian experimental psychology was not born in university laboratories and research institutes, but on clinics, at schools, where its application was a daily necessity. Psychology had no university chair in Hungary till 1918.

The founder and first significant researcher of Hungarian experimental psychology was Pál Ranschburg /1870-1945/.
His name is well-known for homogeneous inhibition which was named after him. He was doing his research work together with education of the handicapped and his healing work on clinics. His results, however, have not remained within the walls of schools and hospitals; his experiments were repeated by H. Münsterberg, the great propagator of psychotechnology and economical psychology on Harvard University. These experiments have significantly supported his explanations for monotony.

The other outstanding personality of Hungarian experimental psychology was Géza Révész /1878-1955/.
Though most of his work was carried our in Holland, his scientific activity had begun in Hungary.
First, his experiments were connected with sensing sounds; this research determined his sound-psychological and music-psychological works later on. These examinations in connection with sound-sensation have led him to the research of abilities and talent, and they were summarized in "Talent und Genie" issued in 1952.

The third determining personality of Hungarian experimental psychology was László Nagy /1857-1931/. He - as a paedolog - has applied the results of psychological examinations first of all in pedagogy, but he himself has done such research work.
His most significant activity was the organization of psychological laboratories in several teachers' training colleges under university level so that candidates could get acquainted with the results of psychology which can be used in education -
- first of all with the help of their own experience.

Beginning of Hungarian experimental psychology was determined by practice.

Ruth HARRIS

St. John's College, Oxford, Junior Research Fellow

Hysteria, Hypnosis and the Salpêtrière: Jean-Martin Charcot and Suggestion

J.-M. Charcot, the Napoléon des névroses, of the late-nineteenth century captured public imagination in the 1880's with his vivid descriptions of hysteria, showing how the disorder followed a law-like course of nervous agitation and contractures which ultimately erupted into a full-blown hysterical paroxysm as frightening and total as an epileptic fit. His work produced a demonology of nervous disorders, a catalogue of the numerous morbid manifestations which the feminine mind and body could conjure up. From sexual ecstacies and hideous groans, to hysterical contractures and paralyses, he documented with the taxonomic fervour of a natural historian, seeking to show the substratum of femininity and the dangerous qualities lurking just beneath the civilised surface. Moreover, he added to his precision with the apparent realism produced by the photographs included in the pages of the Archives de l'iconographie de la Salpêtrière. More than any single work or case study, these images give tangible proof of his belief that the lack of inhibition produced by a feeble and immoral character spawned women more like raving animals than the subdued, docile or nurturing creatures they were meant to be. It was into this clinical milieu that Sigmund Freud was introduced when he came to study in Paris.

Charcot's syndrome of grande hystérie did not last long after his death and indeed was under attack as early as 1882, most notably by Hippolyte Bernheim of the Ecole de Nancy. Bernheim adamantly maintained that Charcot had created a fictitious 'hystérie de culture,' unwittingly suggesting the hysterical symptoms to his subjects, with both physician and patient eliciting ever more extravagant reactions from each other. The point of this paper is not to heap further ignominy on one of the greatest Parisian clinicians, nor to demonstrate how Bernheim's analysis was correct. Rather, it is to investigate the intimate history of the doctor-patient relationship at the Salpêtrière and the constitution of the hysterical diagnosis in fin-de-siècle Paris. Following on from the work of Didi-Huberman and Gérard Wajeman, I intend to examine this process of mutual suggestion, using photographs, records of murder trials in which hysterical subjects attempted to kill psychiatrists, as well as numerous case studies. More importantly, I hope to show the way in which many of the models for the hysterical syndrome were influenced by notions of religious ecstasy and were consistent with Charcot's rabid anti-clericalist stance. Above all, hysteria demonstrated how the martyrs and saints of past ages, as well as present pretenders, were nothing more than neurotics who could now be treated and even sometimes cured of their illusions.

Michael Heidelberger, Freie Universität Berlin, Berlin

FECHNER'S PSYCHOPHYSICS AND THE PROBLEM OF MEASUREMENT

In the following paper I shall try to show how the idea of measurement changed in the course of the debate over psychophysics. I also show that this change contributed to our present notion of measurement. It did not, however influence very much the course of experimental psychology in Germany.

In the 2nd half of the 19th century, a blending of the mechanistic outlook with Kantian philosophy became the dominant philosophy of science in Germany. This was occasioned especially by Helmholtz'sense physiology. Fechner's philosophy was opposed to this "physiological Kantianism".

The conflict between both sides can most easily be recognised and studied in the debate over measurement in psychophysics. Fechner challenged the received view of measurement on two grounds: First, he claimed that the intensity of sensations can be measured fundamentally and second, that the counting of relative frequencies has to be recognised as a central method in the study of man. The first claim was usually countered by pointing out that qualities are only measurable if they can be reduced to space, time or mass. This reduction does not make sense in the case of sensations. The second claim received the objection that statistical methods are good for minimizing errors but not for the measurement of objective attributes or the discovery of causal laws.

Fechner's ideas were received in different ways. G.E.Müller tried to give them a purely physical meaning whereas Wundt turned away from measurement anddefined an experiment mainly as a qualitative analysis of psychic processes. The Neokantians rejected the naturalisation of the ego and the consciousness altogether. Ernst Mach turned the tables by redefining physical measurement in a Fechnerian spirit: physics can be based on sensations and measurement is the numerical representation of the order of our sensations. The psychologist G.F.Lipps wanted to use a similar idea in reforming psychophysics.

In the (Neo-)Kantian philosophy of science measurement is the process by which we make our subjective sensations objective. At the end of the psychophysics debate measurement is the assignment of "inventory numbers"(Mach) which reflect the structure of qualitative relationships between phenomena.

Ingemar Nilsson

Ass Prof, History of Science and Ideas, Univ of Gothenburg

FROM WEIMAR TO MUNICH: POUL BJERRE IN THE PSYCHOANALYTIC MOVEMENT (1911-1913)

Poul Bjerre (1876-1964) were in medical practice in Stockholm from 1907. He used hypnotherapy, a method he had learned from the well-known hypnotist Dr Otto G. Wetterstrand (1833-1907). But he also heard of the new ideas of Sigmund Freud and became curious about their validity. In January 1911 he visited Freud in Vienna at the same time as Alfred Adler began to leave the Freudian circle. Bjerre found Freud unattractive as a person and most of the adherents in Vienna were detestable to him, with the exception of Adler. However, he found so much of value in the Freudian methods and theories that he decided to join the Psycho-Analytical Association. He regarded himself as an independent and highly original member. Freud on his side was impressed by Bjerre's sincerity.
Bjerre sent Freud a paper on a case of chronic paranoia that was included in Jahrbuch für psychoanalytische und psychopathologische Forschungen. It was interesting but unconvincing, Freud wrote to Jung. Jung as an editor tried in vain to persuade Bjerre to give the case a more psychoanalytical explanation. But Bjerre refused to support Freud's theory that paranoia is related to repressed homosexuality. The paper deviated on several points from Freudian theory and practice.
At the psychoanalytical congress in Weimar in September 1911 Bjerre appeared together with Lou Andreas-Salomé, to whom Bjerre had introduced psychoanalysis. The next year she broke their relation and wrote an analysis in her diary of Bjerre's neurotic-obsessive personality.
After the congress Bjerre corresponded with Adler who tried to secure him on his own side against Freud. But Bjerre did not want to subordinate to anyone and he began to work out a system of his own, later named psychosynthesis. At the Munich Congress in 1913 he allied himself with Jung and made a definite break with Freud. He gave a paper on "Consciousness versus unconsciousness" where he confronted himself with Freud's views and gave the therapist a more active role in the analytical process. Bjerre stayed as a member in the society to 1919 Bjerre's relation to Freud and his followers cast interesting light on the formation and boundaries of the psychoanalytical movement. Some of the problems in connection with this will be dealt with in the paper.

Mitchell G. Ash

Assistant Professor, Department of History, University of Iowa

PSYCHOLOGY IN THE "INTELLECTUAL MIGRATION": THE CASE OF KURT LEWIN

Ideas from the social history of science, such as the notion of historically conditioned "national professional styles" in science, are applied here to show how Kurt Lewin's interaction with new institutional and intellectual situations after 1933 affected both his own development and that of psychology in the United States.

The discussion begins with Lewin's intellectual formation before 1933. Under Carl Stumpf at the University of Berlin, Lewin was trained to regard psychological research as a phenomenological prolegomena to the solution of philosophical problems. At the same time, Ernst Cassirer challenged him to consider the metatheoretical issues raised by such an approach. In the 1920s, Lewin tried to weave these two strands of thinking into a coherent unity, undergirding original research with his Berlin students on the psychology of will and affect with a pluralistic philosophy of science. He also showed interest in reforming Tayloristic scientific management in the interest of workers' control, in progressive education and in psychoanalysis. However, Lewin's reception in the United States before 1933 focussed selectively upon his work as a developmental psychologist. This interest led to contacts with influential psychologists and with the Rockefeller Foundation, which became important after 1933.

Lewin's transfer to the United States with the help of these contacts led to important shifts in his institutional situation and in his thinking. Almost alone among emigre psychologists, Lewin was able to establish alternative settings outside normal disciplinary channels for the discussion of his ideas. At the same time, he linked up with a developing network of private, semi-private and public organisations financing research in the social sciences, and with the liberal wing of the psychological profession in the Society for the Psychological Study of Social Issues. These activities were intertwined with the development of Lewin's thought from a "dynamic theory of personality" to a general field-theoretical approach, including social psychology. This was rooted in part in his confrontation with American educational practices and in his own situation as a German Jew in exile. Finally, the paper examines the selective integration of Lewinian ideas and techniques after 1945 in light of American psychology's development into specialized sub-disciplines with different methodological and theoretical preferences.

Lewin's career in the United States was a success story. But the shaping of that success reveals the complex interaction between his biographical "processing" of the emigre experience, the intellectual dialogue of emigres with their new environment, and the selection Americans made from the results of these processes.

Helio Carpintero

University of Valencia, Spain   Professor of Psychology

The development of psychology in Spain: A sociobibliometric approach

The development of scientific psychology in Spain is presented by means of a bibliometric approach based on the writings of important authors. It reveals their intellectual backgrounds as well as the sociopolitical factors affecting their works, such the Spanish Civil War (1936-1939).

Mr. Bertrand PULMAN

Administrateur du Collège International de Philosophie, Paris, France.

The dawn of the Science of Religion.

This lecture will present the results of a research on the causes and mechanisms which presided over the institutional and discursive emergence of a "Science of Religion", in Europe, between 1860 and 1880 (Emile Burnouf, La Science des Religions, Paris, 1870; Max Müller, Introduction to the Science of Religion, London, 1873).

Three hypothesises will be developped:

- The birth of the Science of Religion must be considered as a consequence of the formation of a philological positivity at the beginning of the 19th century. Following the direction of Michel Foucault's works (Les mots et les choses, Paris, 1966), we will show that the intelligibility of the religious subject has been modified according to the upheavals which have affected, through philology, the status of signs in the 19th century.

- The influence of philology on the birth of the Science of Religion has conveyed a central role to the historical method. Instead of repeating the "already said" of the religious Text, the exegesis of scientists involved a critical and desacralizing method, by analysing the forms and history of the "said".

- This movement arose with the study of Sanskrit in the field of Oriental religions (Chézy, Burnouf); it has then spread, through Hebraic studies (Renan), in the direction of Biblical investigations. Therefore, according to a mechanism essential to its constitution, the Science of Religion has been elaborated through a preliminary reflexion on Oriental religions; the scientific knowledge by the Western World of its own religious tradition could only establish itself by means of a discourse on what was most foreign to us.

Bernard-Pierre LECUYER

Maître de Recherche au Centre National de la Recherche Scientifique (Paris, France)

Vital and social statistics in France from the Old Regime to the Third Republic : Changes and continuity.

In opposition to the contrasted characteristics of the political history of France between the French Revolution and the Third Republic, the history of vital and social statistics, although partly responsive to the frequent political upheavals, must also be seen as a regular and almost continuous development.

On the surface, the administrative machinery in charge of collecting, assembling, and, later, publishing the data, has undergone many transformations. Thus, the "Contrôle général des finances", primarily responsible for collecting demographic and economic statistics under the Old Regime, was de facto abolished in 1789 and revived only in 1800 with the "Bureau de statistique" created by Lucien Bonaparte. Again this bureau was suppressed in 1811 (when the situation of the Great Empire began to be shaken) and ressuscitated again only in 1834 by Thiers as "Statistique générale de la France". To complete the story the "Statistique générale de la France" was replaced in 1945 by the present "Institut National de la Statistique et des Etudes Economiques" a branch of the Ministry of Finances.

If one looks closer at the work actually performed, and the personnel employed, one finds even during the most apparently disruptive periods a much greater amount of continuity. For example, the census-type operations performed in 1801 and 1806 had been preceded by repeated attempts during the Old Regime and even the Revolution (François de Neufchâteau completed one such attempt under the Directorate). It is also meaningful to notice that the "Recherches statistiques sur la Ville de Paris..." undertaken in 1817 under the auspices of Prefect Chabrol with the advice of the former Napoleonic Prefect Baron Fourier was partly the execution of an exhaustive survey of the generality of Paris which had been planned by the then Intendant of Paris Bertier de Sauvigny.

Since the "invention" of the mortality table (Dupâquier J. and M., Histoire de la démographie, 1985), the problematics of population studies was established in its broader lines. Nevertheless, the introduction of new categories or entries in French official statistics did happen and is thus worth being studied.

Hervé Dumez

Centre de Recherche en Gestion. Ecole Polytechnique. Paris.

The emergence of mathematical economics: the case Léon Walras.

The emergence in France of the mathematical conception of economic phenomena by Walras can be hold as a symptom of a multidimensional change in the theoretical economic field. Before, political economy was mainly ideology: being an economist was not a profession; economics had to be taught to pupils and labour classes, just as morals; and son on. The emergence of the walrasian economic theory coincides with deep changes in the field. Since the 1870-1900's (the "walrasian era"), a new economic theory has to comply with three kinds of demands:

-a scientific and academic demand. As the economist has become a professionnal (1885: creation of the American Economic Association), a new theory has to work on unanswered questions and unresolved issues. It has also to supply "puzzles" (T.S. Kuhn), topics for researchers, and son on.

-an ideological demand. Nowadays, this demand is above all negative: an economist is prevented from proposing economic decisions which could be inconsistent with the existing organisation of society.

-a technical and managerial demand. An economist is expected to provide the government (macro-economics) and the manager (micro-economics) with decision making models. This kind of demand appeared with the "walrasian era".

(this paper is an abstract from a book to be published -Dumez (Hervé): L'Economiste, la Science et le Pouvoir: le cas Walras. Paris, Presses Universitaires de France, 1985)

Henrika Kuklick

University of Pennsylvania, Philadelphia, PA, U.S.A.

Evolutionism in Early American Social Science

Nineteenth century American social scientists were typically Lamarckian evolutionists. Their theoretical framework joined all varieties of social scientists in related, frequently indistinguishable enterprises. Clear disciplinary differentiation was not evident until the turn of the twentieth century. It was the product of three factors: advances in knowledge in the biological sciences; the organizational dynamic of academic institutionalization; changes in cultural attitudes. In accounting for intellectual change in this paper I focus on the first factor.

For the differentiating social sciences adopted various research programs that can be described as different responses to newly-available information about the processes of inheritance. On the grounds that biological and cultural processes had been shown to be altogether unrelated, anthropology repudiated the goal of describing social organization and change in evolutionary terms. Though in practice psychology ignored questions of evolutionary adaptation, psychology's emergent research program was justified by appeal to recent advances in understanding of biological evolution. Sociology remained evolutionist but ceased to be Lamarckist, its research program inspired by ecological metaphors. Economics increasingly turned toward formalist theory, its accounts premised on static models of human nature.

George W. Stocking, Jr.

Director, The Morris Fishbein Center for the Study of the History of Science and Medicine, The University of Chicago

THE CRISIS IN BRITISH EVOLUTIONARY ANTHROPOLOGY, 1900-1910

Although it is well known that twentieth-century anthropology emerged in the reaction against social and cultural evolutionism, the details of that process have been better studied in the United States than in Great Britain. We know a good bit about later American evolutionists and the early work of Franz Boas, but despite Ian Langham's work on W. H. R. Rivers, we still know relatively little about the history of British anthropology between the grand period of classical evolutionism (c. 1870) and the simultaneous publication in 1922 of major works by the two founding figures of the modern functionalist tradition, Bronislaw Malinowski and A. R. Radcliffe-Brown. Specifically, the debate about the problem of totemism provoked in large part by the second edition of James G. Frazer's <u>Golden Bough</u> has been little investigated. This paper will focus on the work of two of Frazer's critics: the literary anthropologist Andrew Lang, a disciple of E. B. Tylor who had begun to question evolutionary intellectualism in the 1890s, and who wrote on various aspects of social organization after 1900; and R. R. Marett, Tylor's successor as Reader of Anthropology at Oxford, who wrote a series of very influential papers on "preanimistic religion" between 1900 and 1908. By 1910, empirical anomaly and conceptual confusion had precipitated a crisis in the evolutionary paradigm--well exemplified in Frazer's four volume <u>Totemism and Exogamy</u> (1912). It is in this context, as well as that of Rivers' influence that the early work of Malinowski and Radcliffe-Brown must be interpreted.

Mary Morgan

Lecturer, University of York, York, England

MATHEMATICAL MODELS AND STATISTICAL RELATIONSHIPS IN THE HISTORY OF ECONOMICS

Mathematical and statistical methods invaded economics during the first half of the twentieth century. Two distinct schools of thought governed the role of quantitative methods in economics and this in turn dictated the methods developed and practised in each.

The "econometrics" school took the union of mathematics and statistics with economics as their ideal. In this ideal, mathematics were to make deductive economic reasoning more rigorous while statistics were to measure and test economic laws in substitute for the experimental method. In the early years of econometrics, the methods of mathematics and statistics were indeed complementary - econometricians specialised both in deriving mathematical representations of economic theories, in developing the dynamic schemes required to mould economic models to the demands of observed economic data and in the development and application of appropriate statistical techniques to measure and test these mathematical models.

The alternative "quantitative" school of economists adopted statistical and arithmetical techniques to enrich a descriptive, hypothesis-seeking approach to economics which was blatently empirical in outlook. The intended role of quantitative methods was to reveal the laws of economics which lay hidden in the data. Although based on the same initial set of statistical tools, the methods evolved by the "quantitative" school were very different from those developed by the "econometric" school.

Michael Pollak, Institut d' Histoire du Temps Présent,
    CNRS, Paris

LA FONDATION FORD ET LES SCIENCES SOCIALES EUROPEENNES

   Dans l' attente d' une augmentation rapide des
moyens financiers à sa disposition, la Fondation Ford
a fait préparer en 1949 un rapport détaillé sur les
champs d' action qui est devenu une sorte de programme-
cadre. Ce programme très ambitieux se propose de contri-
buer à la paix internationale et au renforcement de la
démocratie. A cet effet, le rapport distingue cinq domaines d' action: la paix, le renforcement de la démocratie, celui de l' économie, l' éducation, l' étude du com-
portement individuel et des relations humaines. Ce programme vise à préparer les Etats-Unis à jouer le rôle
prépondérant dans le monde qui leur incombe à la sortie
de la Deuxième Guerre Mondiale (R. Gaither, et al.,
Report of the Study for the Ford Foundation on Policy
and Program, Detroit, Michigan, Ford Foundation, 1949,
vol.I).
   Dans le tome 3 sur les sciences sociales, celles-
ci sont présentées comme des sciences anhistoriques des
attitudes, des comportements et de l' organisation. Par-
tant de la même démarche que les sciences exactes, les
sciences sociales accusent un retard qu' il s' agit de
combler. Grâce au développement d' une telle approche,
on pourrait, toujours selon ce rapport, mieux comprendre
et gérer le changement social.
   Appliquée à l' Europe, cette politique des sciences
sociales devait favoriser l' analyse concrète et une
approche politique pragmatique en rupture avec les pen-
chants globalisants et les traditions marxistes. Suivre
dans une approche comparative la mise en oeuvre de cette
conception dans deux contextes nationaux différents,
la France (Ford Foundation Archives, PA 58-37, PA 60-437,
PA 61-147) et la République Fédérale d' Allemagne (Ford
Foundation Archives, PA 56-22, PA 58-260, PA 55-217) permet de montrer comment celle-ci a effectivement contri-
bué à réorienter les sciences sociales européennes, mais
également les modifications des intentions de cette politique en fonction des enjeux intellectuels des différents
contextes d' accueil tout au long de sa réalisation.

DESPOIX Philippe

Student   (Ph. D. EHESS Paris, FU Berlin)

Affinity of the new hungarian and czechoslovak sociologies

The new hungarian and czechoslovak sociologies first appeared in the early sixties as a separate field of research in the context of destalinization. Through a break with pre-war sociological traditions, the developpement of sociology connects with the necessity of reforming and modernizing the economy, of building up a new system of social management. This official framework faced their main trends with the dilemma of the contradiction between a real critical analysis and a goal-directed manipulation of the social data of evolution. An appraisal of the historical functional relevance of these new sociologies could distinguish two main directions which cut across the developpement of both countries, despite the differences due to the separate traditions. First and most important, the reform sociology, which concentrated on the social consequences of the technical and scientific revolution, saw the main problem of modernization in the possible contradiction between pure economical optimalization and the program of social humanization (Richta and Klein, or Hegedüs..). The second trend, critical sociology, tried to resolve this antinomy in the search for an optimal model, which it saw in a general principle of self-management including all levels of society (economy, political sphere and culture). But their basic conceptions are, for both trends, dependent on other theoritical frameworks (concepts of needs, norms and values..) of social science and philosophy, mainly that of critical marxism, that of social-ontology, and phenomenology.

Jozef Babicz

Polska Akademia Nauk, Instytut Historii Nauki

The geographical scientific schools:
systematic and historical approach.

Recent papers presented at the Symposium on the History of Geographic Thought indicated several types, personal and national schools. Among factors discussed were internationalization, socialization, and diversification (Pinchemel); professionalization and institutionalization (Berdoulay); their significance for the development of the discipline (Claval). The role of historical context for national schools considered as historical phenomena of a given epoch was also discussed (Hooson). The author confronts the above picture with two collective publications: 1) Schools in Science (Moscow 1977), representative of Soviet views and 2) Schools in Science (Warsaw, 1981) based on Anglo-American literature studied by Polish authors. On this basis, the author suggests enlarging the concept of geographical schools by defining exactly their additional types and forms in order to stimulate further historical studies.

Ge Jianxiong, Fudan University, Shanghai, China

OF THE MAKING OF THE CHINESE HISTORICAL MAPS

China has a long history of historical geography studies and the production of historical maps. Pei Xiu (裴秀), one of the acknowledge map-makers in antient China, fomulated his Zhi Tu Liu Ti (制图六体, six methods of charting) and produced Yu Gong Regional Maps (禹贡地域图) in the third century A.D.. Employed the method of Gu Mo Jin Zhu (古墨今朱, the historical in black ink while the contemporary in red ink), Jia Dan (贾耽) drew his Hai Nei Hua Yi Tu (海内华夷图, Maps of the Hana and National Minorities of China) which still exists two reduced-size reproductions carved on stone in 1137 A.D.. About 900 years ago Shui An Li (税安礼) produced his Li Dai Di Li Zhi Zhang Tu (历代地理指掌图, Chronological Directory Maps of Historical Geography), 44 maps in all, that "records historical changes and makes comparisons". Wooden-plate reproductions of Shiu's maps are now available. And we have today more than 10 wooden-plate printed atlases of historical geography that came out between Shiu's time and the end of Qing (清) Dynasty. At the turn of the present century, Yang Shou Jing (杨守敬) with his students compiled, drew and eventually published his Li Dai Yu Ti Tu (历代舆地图, The Chronological Maps of Historical Geography), a book of 34 string-bound volumes which covers the period from the 8th century B.C. down to the 17th century A.D.. This brilliant work achieved a milestone in the history of historical map making, notably for its unprecedentedly exhaustive details. The Historical Atlas of China, with Prof. Tan Qi Xiang as its chief-editor, devides into 8 volumes with 20 map-groups from the stone age to the end of last century that hold 304 maps and about 70,000 historical place names. This collection of maps is to be counted as a monumental work unprecedented in China as well as in the world.

Horacio CAPEL, Universidad de Barcelona, Spain

RELIGIOUS BELIEFS, PHILOSOPHY AND SCIENTIFIC THEORY IN THE ORIGIN OF SPANISH GEOMORPHOLOGY. XVII-XVIII CENTURIES.

At the end of the XVII century, the works of Newton culminated a century of deep transformations in the field of physics and astronomy, which led way to modern science. During those same years naturalists fervently debated such topics as the location of Paradise on Earth, geography before the Great Flood and the extension and consequences of the Flood, or the possibility that the deposited - petrified seashells were formed by the scallops of the pilgrims. The more audacious dedicated their time to elaborate sacred physics or a sacred theory of the Earth so as to try to explain, with scientific arguments, the story of the creation of the world transmitted by Genesis. Everyone accepted that Earth was created by God as Man's dwelling, that it was about 6.000 years old and that, since its creation, it had not undergone any further changes than those referred to in the sacred books. A century later, at the end of the XVIII century, scientists admitted that our planet had a History, and that it could even be millions of years old.

This paper attempts to cancel certain circumstances that contribute to explaining the delay in the development of modern geology and, at the same time, to determine the terms that produced the debate going from one anthropocentric, teleological and providentialst concept, to another that accepts the idea of changes and evolution, that rejects finalism, and that turns to the laws of physics to explain the history and structure of the Earth.

In the transition towards scientific geology, the theological debate, the philological and historical discussions and the results of empirical observation seem inseparably associated. The empirical and erudite data gathered together concerning erosion, the transportation and depositing of materials, as well as the discovery of petrified organic remains, could not be correctly interpreted until there were new and adaquate conceptual frameworks. These began to take form through disputes concerning such polemic topics as the relative importance the old and the new, the location of Paradise on Earth, the meaning and scope of God's punishment on Man, the mechanism through which the Great Flood came about and its extension to the American continent, the existence of giants and long-lived or macrobiotic men in the past, the origin of the American Indians, the existence of preadamites or the providence concept, and the possibilities of miracles in a world ruled by natural laws.

D. R. Stoddart

Department of Geography, Cambridge University, Cambridge, England

INTERPRETERS AND INTERPRETATIONS IN THE EARTH SCIENCES: THE SIGNIFICANCE OF FREUD

Of the dominant intellectual figures of late nineteenth century European thought, the influence of Darwin and Marx on Geography and cognate sciences has been exhaustively analysed, but that of Freud almost completely ignored.

This paper discusses the influence of Freud on Geography and Geomorphology in the period 1900-1940. It discusses the symbolism of geographical features and their Freudian interpretation, with particular reference to rivers and mountains. It is also concerned with the understanding of scientists and their work through Freudian concepts, and the ways in which this affected their mode of argument and choice of subject. Particular attention is paid to the work of Arthur Tansley and C. C. Fagg, both of whom had direct contact with Freud, and also to the frontier hypothesis of Frederick Jackson Turner.

Conclusions specific to the earth sciences are then related to the wider cultural context of European thought.

DORY Daniel

Chercheur, U.A. 04 0914 CNRS-UNIVERSITE PARIS I

Histoire de la connaissance d'un territoire ethnique africain: le Lobi (1900 - 1985).

Plutôt que de l'histoire d'une discipline à un moment donné, cette communication traitera des **visions successives** d'un territoire ethnique de l'Afrique Occidentale,(N-E de la Côte d'Ivoire et S-O du Burkina Faso; le Lobi ghanéen ne sera pas envisagé ici).

Après une première serie de travaux provenant essentiellement des militaires engagés dans la conquête (1900-1920), on assiste, en un second temps à l'émergance des premières synthèses (1921-1935) tant ethnosociologiques que naturalistes, dont les auteurs ont avant tout pour souci de pacifier et rentabiliser le territoire placé en situation coloniale.

Suit une période de relatif effacement des recherches concernant le Lobi, (1936-1970),dû en grande partie au fait que le territoire se révéla moins riche que prévu, (peu d'or notamment),et que par conséquent les efforts d'inventaire devinrent moins urgents.

Enfin, depuis une quinzaine d'années on voit à nouveau une multiplication des travaux menés dans le champ de sciences très diverses (anthropologie, géographie, linguistique, pédologie, etc.), ce qui est à mettre en rapport tant avec le développement de ces domaines particuliers qu'avec les diverses péripéties des politiques de modernisation entreprises au Lobi au cours des dernières années.

Il s'agira donc de tenter d'élaborer une histoire de l'émergence de savoirs centrés sur un objet complexe particulier en fonction de conditions contextuelles changeantes.

Berardo Cori, Università di Pisa, Pisa, Italy

INNOVATIVE FERMENTS AND CURBING MECHANISMS IN ITALIAN
CONTEMPORARY GEOGRAPHY

In the Italian geography there are now new forces: even if
they still lack a true group of renewers, a number of talented
and original thinkers, born in the 1930s, has reached university chairs, and another unit of younger men, born in the 1940s,
yet equally valid, is advancing as reinforcement.

There are now, to a greater extent than in the past, means
of attenuating what has euphemistically been called the gap in
cultural interests between the Italian and international scientific communities: today, Italian geographers travel more, read
more books and journals from abroad, take advantage of the scientific consultancy of authoritative foreign geographers and
have a considerable number of works in translation available,
which, now found extensively in student reading lists, compensate for the limitations and inadequacies of Italian original
production.

The work of demystification carried on by geographers such
as Lucio Gambi, the new forces, the revivified contribution of
the best exponents of previous generations such as Aldo Sestini, and new instruments are already having a certain positive
effect under our very eyes. Some works of synthesis which appeared in the late 1970s have resumed the tradition of 'major
works' which seemed to disappear with Umberto Toschi. Other important works are in advanced stage of preparation, from a universal geography, extensively revised in content, to a national thematic atlas which will finally fill a well-noted gap. Greater involvement of geographers is being seen in the study of
problems of planning, the Mezzogiorno, regions, ecology, protection of cultural and environmental wealth, as well as in solid participation in initiatives promoted by state bodies and
local authorities. Interdisciplinary dialogue seems to have become free of inferiority complexes. Theoretical work is becoming more frequent and, what matters most, more pithy, whilst
the epistemological debate between the various tendencies, if
still a little long-winded, has reached a high level. Finally
and above all, there is a breakthrough into Italy, with reasoned application and discussion, of the most advanced geographical themes and methodologies.

But some curbing elements are working against these innovative ferments, acting by means of the unyielding mechanism of
competitive examination self-reproduction. They are the conformism, some harmful aspects of individualism, the cultural backwardness, the narrow-mindedness, the dogmatism, the methodological rigidity, the fossilization, the lack of self-criticism
and renewal, the cultural pecking-order based on age and academic status, and finally the anathemas — this isn't geography!

Vincent Berdoulay

Département de géographie, Université d'Ottawa, Ottawa, Canada

Les difficultés de la géographie dues à l'avatar positiviste de la classification des sciences

Le contexte institutionnel dans lequel la géographie doit se développer dépend toujours dans une certaine mesure des conceptions implicites ou explicites que ses responsables ont de la division du travail scientifique. La répartition des efforts en départements, instituts ou facultés, le traitement de l'information documentaire, les organismes subventionnaires, etc. exercent une pression, fréquemment négative, sur la constitution d'approches disciplinaires intégrées. C'est ainsi souvent le cas de la géographie. Or le poids d'une classification établie il y a un siècle et demi par Auguste Comte se fait toujours sentir dans de nombreux pays, par l'intermédiaire des gestionnaires de ces institutions et des scientifiques qui y collaborent. Il s'agit d'une conception "populaire" (par opposition à une conception plus au fait de la recherche) dont la force doit nous inciter à en faire la critique.

On se penchera donc d'abord sur la classification proposée par Auguste Comte. Ses principes fondamentaux et leur résultat seront exposés. Ils sont en effet à la base des classifications positivistes qui perdurent jusqu'à ce jour. Celles-ci ont en commun un caractère linéaire, que l'on retrouve même chez les critiques de Comte, tels que Herbert Spencer. D'autres classifications issues du néo-positivisme seront aussi examinées. On verra encore que la recherche géographique qui veut en tenir compte doit soit se glisser dans un lit de Procuste extrêmement réducteur, soit se contenter de projeter l'image d'une géographie comme science concrète non-théorique, sans grande respectabilité scientifique.

Les géographes ont donc intérêt à dénoncer les classifications positivistes dérivées plus ou moins inconsciemment de la pensée de Comte. Ils doivent se tourner vers d'autres justifications - non linéaires - des rapports entre les sciences.

Graham Rees, The Polytechnic, Wolverhampton, England.

FRANCIS BACON, RELIGION AND THEORY SELECTION

Francis Bacon believed that natural philosophy had no contribution to make to discourse about the mysteries of the faith and that theological talk should be excluded from natural philosophy. Human reason was incapable of dealing with the revealed mysteries, and theology was (on the whole) to be kept out of the sciences.

But while science was not subject to <u>undue</u> circumscrip-ion, it was nevertheless to be <u>limited</u> by religion. In practice this meant that theological presuppositions should not necessarily be excluded from the processes of theory selection or formation, even though <u>subsequent</u> justification of a theory would be conducted without explicit recourse to theology. Theology placed limits on choice but had no part to play once a choice had been made. Natural philosophy should not be <u>invaded</u> by revealed theology, but is nevertheless an activity <u>bounded</u> by theological truths.

Bacon's understanding of the revealed truth affected his attitudes to a wide range of topics, but above all it affected his attitude to systems of theories. Systems were to be judged by the extent to which they implied or embodied the "facts" of Holy Writ - especially "facts" about the Creation. On these grounds the philosophies of Telesio and the atomists were found wanting. Bacon however chose to base his own vast system of substantive science on Paracelsian ideas, ideas which (according to their originators) were full of Scriptural virtue. But though Bacon chose to work from Paracelsian models, and so produced a system which he believed conformed to Holy Writ, one looks in vain for direct references to Genesis or attempts to legitimise the system by representing it as infallibly rooted in Scripture. He excluded Holy Writ from any role in validating theories once they had been selected, and rebuked the Paracelsians for trying to hijack the Bible: the requirement that a system should conform to the Text was quite different from the claim that a system was the one true explication of the theolog-ical "data".

Examination of this and related topics enables us to understand why Bacon avoided seeking open theological support for his own theories; more significantly it helps us to grasp one of the most important principles of his eclecticism: that theological truths acted as boundary conditions for his philosophical choices.

Margaret J. Osler

Associate Professor of History, University of Calgary, Calgary, Alberta, Canada

The Theological Foundations of the Mechanical Philosophy

René Descartes (1590-1650) and Pierre Gassendi (1592-1655) both espoused the mechanical philosophy, the main tenet of which was that all natural phenomena should be explained in terms of matter and motion. Their views were very influential on science during the latter decades of the 17th century. Whereas they agreed on the most fundamental level, they disagreed about the nature of matter and on their ideas about scientific method. I have traced their differences to their underlying assumptions about God's relationship to the creation. Gassendi adopted a voluntarist theology, according to which emphasis is placed on God's will and his ability to intervene in the natural order at any time. Hence, laws of nature, according to Gassendi, are merely empirical generalizations of what we observe as regularities in nature.

This paper will deal with the acrimonious interchange between the two founding fathers of the mechanical philosophy, an interchange that occured following the publications of Descartes' <u>Meditations</u> in 1641. I will argue that their debate and the particular issues that divided them can best be understood in terms of the theological assumptions that each made. In this sense, their disagreement paralleled the famous Leibniz-Clarke debate of 1717, which rested ultimately on the same theological differences. The fundamental aim of this paper is to show how different styles of science came into being as a result of differences in philosophies of nature which were rooted in different theological coneptions about God and nature.

Fritz Krafft

Johannes Gutenberg-Universität Mainz, Federal Republic of Germany

Johannes Kepler: Astronomy as a Way of Worship

The advance of modern science by no means presents itself as being exclusively determined by the degree of its 'emancipation' from theology and religion, as it has been maintained again and again echoing 19th century traditions. On the contrary, recognizing nature was understood also as reflecting on and comprehension of the divine plan of Creation. Indeed, it was just this element which initiated the great innovative thrust within the thousands of years of astronomy which helped it to overcome the way of thinking of antiquity and the Middle Ages. In the first instance, it was the objective of Johannes Kepler's scientific lifework to discover the divine harmony of the cosmos as the reason for number, size and movement of the planets.

This is expressed as the impetus and aim of his studies not only in one of his last works, the 'Harmonice mundi', but already in the foreword of his first work, the 'Mysterium cosmographicum' (1596). However, for him this harmony was not limited to mathematical proportions (as in the case of Plato and the Pythagoreans); for him mathematical proportions were only an expression of the divine will of Creation, as principles of order in a _natural_ world which were observed in a _natural_ (physical) way. In other words, Kepler wanted to make (mathematical and exclusively empirical) astronomy into harmonics, as well as into physics again (that means also harmonics and physics into empirical sciences). Up to that time all three disciplines were separated from and conflicting with each other; he wanted to combine them into one synthesis. Kepler's lifework makes it clear that his conviction, that the synthesis of all three approaches was necessary, was the presumption also for those discoveries which proved to be correct beyond the original context, namely the three Keplerian Laws which broke with the old demand of the astronomers that all celestial motions had to be described by uniform circular movements.

Gary B. Deason

Associate Professor of History and Religion, St. Olaf College,
Northfield, Minnesota U.S.A.

THE REFORMATION AND SCIENCE

　　　　The literature of the Reformation and early modern science has produced what may be called a strong interpretation of their relation. Merton, Hooykaas, Mason, and Klaaren hold that Protestant doctrines or values contributed directly to the rise of science. This paper does not deny a contribution of Protestantism to science, but qualifies the strong interpretation by showing that it does not apply, in its present form, in two important cases: (1) Bacon's use of Protestant concepts and language to argue for inductive methods and (2) Boyle's and Newton's use of Protestant ideas to argue for the passivity of matter. In these cases, the paper states, the Reformation tradition did not inspire scientific work, but provided a source for constructing arguments favoring the new science against Aristotle. This use gave Protestantism a supporting role in debates about science, but one that should be distinguished from the inspirational role in scientific investigation attributed to it by the strong interpretation.

John Henry, Research Fellow,
Wellcome Institute for the History of Medicine, London

Henry More and the Spirit of Nature

　　Henry More's concept of the Spirit of Nature, a hylarchic principle which he believed to be essential to account for 'such Phaenomena in the world, by directing the parts of Matter and their Motion, as cannot be resolved into mere mechanical powers'(Collection of Several Philosophical Writings, 1662, vol.II,p. 193), is present in his earliest philosophical works and received its fullest treatment in his Immortality of the Soul of 1659. Throughout the 1660s More and his neo-Platonizing disciple, Joseph Glanvill, were regarded as allies of the new philosophy. In 1671, however, when More reiterated his views on the Spirit of Nature in his Enchiridion metaphysicum, this alliance rapidly broke down. Although More's basic concept had not changed since 1659, the way in which he presented his ideas gave them a new rhetorical impact. His rhetoric was now deeply disturbing to leading representatives of the new philosophy, Boyle, Hooke and Oldenburg,and to scientific virtuosi like John Beale and Sir Matthew Hale. For the first time these thinkers sought to dissociate themselves from More's views and to refute his arguments. the aim of this paper is to explain the circumstances surrounding More's changing attitude to the new science and the mechanical philosophers' new awareness that More's philosophy must be repudiated.

Richard H. Popkin

Professor of Philosophy, Washington University, St. Louis,

Sir Isaac Newton and the Rise of Fundamentalism in America and England

Sir Isaac Newton's theological views have not been give much serious attention. Most of his writings on the subject have still not been published.

The two published works of Newton on theology, <u>The Chronology of Ancient Kingdoms</u>, and <u>Observations on the Prophecies in the Book of Daniel and The Revelation of St. John</u>, were carefully studied by English and American Millenarians. At the close of the 18th century, some theologians in America, Elias Boudinot and Charles Crawford, and the English theologian, George Stanley Faber, used ideas from Newton in order to formulate the principles that became central for fundamentalism, and to create the institutions that guided its development. Thus Newton's theology played an important role in the rise of Fundamentalism in America and England after the American and French Revolutions.

James E. Force

Associate Professor, Department of Philosophy, University of Kentucky

BEYOND THE DESIGN ARGUMENT: NEWTON'S "SLEEPING ARGUMENT" AND THE NEWTONIAN SYNTHESIS OF SCIENCE AND RELIGION

Given the acknowledged and much-demonstrated importance of the design argument in the eighteenth-century English attempt to link science and religion, the conclusion of Newton's first and most revealing letter to Richard Bentley (December 10, 1692) demands one's attention. After rejoicing about Bentley's utilization of Newton's description of the frame of nature in the Principia to produce an important statement of the design argument, Newton goes on to say that

> There is yet another argument for a Deity wch I take to be a very strong one, but till ye principles on wch tis grounded be better received I think it advisable to let it sleep.

Utilizing evidence drawn from Newton's contemporary, William Whiston, and from Newton's "sleeping argument" is the illustration of God's dominion and Lordship over human history and destiny through properly interpreted prophetic history. Just as the design argument illustrates God's dominion over the "World Natural", the record of fulfilled prophecies reveals God's total mastery and control over the "World Politic". Ultimately, this aspect of the Newtonian synthesis of science and religion must be based upon a reinterpretation of the distinctive manner in which religious, scientific, and political language is integrated in the thought of Newton and his like-minded contemporaries. For them, the Bible is an experimental casebook which affords, if interpreted according to objective and rational "Principles" (and not according to private "fansies" or vain hypotheses), a solid record of God's mastery and control over the human past and the extremely strong probability that this dominion will extend into the future.

Newton's use of the "sleeping argument" in his manuscripts and Whiston's public statements of it to supplement the design argument and to reveal another aspect of God's dominion over creation opens up the possiblity that, for early Newtonian scientist-exegetes, the nature of the English alliance between science and religion extends far beyond the boundaries of the conventional design argument.

Simon Varey

Lecturer in English, University of Utrecht

DIVINE ARCHITECTURE

In this talk I shall discuss the theory and practice of English architecture in the 18th Century, with special reference to the work of John Wood, who designed most of Georgian Bath. Neo-Palladian architecture in England was derived from a peculiarly English Vitruvianism, first expressed by John Dee and Inigo Jones, but not widely recognized until the late 17th Century. Thanks mainly to John Evelyn, Renaissance architectural theory in England at last began to absorb and be absorbed by continental Palladianism by about 1700. The resulting Palladian aesthetic was a melange of literary, scientific, theological, and historical traditions.

I shall describe and interpret Wood's achievement, his conceptions of space, and his response to these traditions. I shall show how he extended current neo-Palladian theory by ascribing the science of architecture—especially its reliance on geometry—to divine revelation. Alone among British architects, Wood was able to put his theory into practice, both in his designs for individual buildings and in his urban planning.

Dr John Hedley Brooke

Senior Lecturer in History of Science, University of Lancaster, England

NATURAL THEOLOGY AND THE SCOPE OF THE SCIENCES : PERSPECTIVES ON THE NINETEENTH-CENTURY DEBATE.

Recent scholarship has exposed the diversity of 19th-century natural theology and, with it, the inadequacy of stereotypes which reduce the Darwinian revolution to an antithesis between naturalistic and theistic explanation. Among its many functions natural theology could itself be a vehicle for extending the domain of natural law. Moreover, the investigation of idealist philosophies of nature, spanning the pre- and post-Darwinian periods has revealed a series of intermediate positions between the positivism (and sometimes materialism) of Darwin's more flamboyant disciples and the images of creation enshrined within traditional christianity. Add a growing sensitivity on the part of historians of science towards the social, political and religious aspects of natural theology, and questions concerning its fate become more complex and engaging.

One object of the paper will be to explore recent refinements to the history of natural theology, some of which can be brought into focus by discussing the relationship between scientific knowledge and religious belief as envisaged by the Cambridge mathematician, philosopher and moralist, William Whewell. Such a discussion requires an analysis of the ethical dimensions of natural theology, an awareness of the non-demonstrative functions of arguments for design, and a recognition of more than one ironic consequence of erecting a religious edifice on the unity and uniformity of nature.

Frederick Gregory, University of Florida, Gainesville, Florida, US

THE RECEPTION OF DARWIN AMONG GERMAN THEOLOGIANS

The German protestant theological community of the latter half of the nineteenth century developed largely in isolation from natural science. At mid century neither the neo-Lutherans of the Erlangen School, nor the <u>Vermittlungstheologen</u> at Berlin and elsewhere, nor the speculative theologians in the lingering heritage of Hegel were much concerned with or threatened by the burgeoning scientific community in Germany. When these three dominant strains of German protestant theology then gave way to Ritschl's attempt to cut theology loose from metaphysics, the irrelevance of natural science for theology seemed clearer than ever.

Even though most German theologians harbored no active interest in natural science, there were a few who did. No development in science forced its way into theology in the nineteenth century more directly than Charles Darwin's theory of evolution by natural selection. Because Darwin's theory vitiated the sharp distinction between the metaphysical and the ethical on which many German theologians relied, it was impossible for the German theological community to ignore Darwin completely. The reception of Darwin among German theologians contrasts markedly from that in other lands. Knowledge of the German theological response contributes to our continuing attempts to assess accurately and to describe properly the historical relations between science and religion in the nineteenth century.

Daniel Gasman

Professor of History, Graduate School, City Univ. of New York

HAECKEL'S RELIGIOUS MONISM, ITS CULTURAL IMPACT

The late nineteenth and early twentieth century witnessed a reaction against materialism and a vigorous revival of mystical, occult, theosophical, and religious writings in Germany and in other European countries. The belief that the 'Riddle' of the universe points to the existence of a higher Reality, the renewed quest for the absolute and the divine, and the search for the organizing principle of the world was largely rooted in and drew sustenance from the evolutionary religious Monism of Ernst Haeckel and the German Darwinist movement which he created. It was Haeckel's biological science, combined with his conception of evolution and religion, his doctrine of the oneness of nature and man and his belief in the unifying existence of the world-soul that stimulated and nourished the widespread cultural attempt to synthesize religious mysticism and science. Yet, Haeckel's reputation as a materialist and mechanistic Darwinist remains largely uncontested in conventional accounts of the period and the assumption is still readily made that Haeckel led the way to the general secularization of society. Haeckel, however, was a materialist more in name and rhetoric than in fact, and his religion of Monism and conception of evolution were radically different from the naturalism and evolutionary science of Darwin, Huxley, and Spencer. Hardly an opponent of religion (only of Christianity), Haeckel was a popular cultural hero, endowed with the prestige of one of Europe's most famous scientists. Therefore, his decidedly religious ideas and programs became a uniquely powerful intellectual movement, affecting not only biology but also the other sciences, as well as the art, literature, philosophy, politics, and society of the period. Haeckel's Monism did not fundamentally express materialism but rather manifested a revolution against materialism and served as a stimulus for the foremost underlying cultural theme of the age, a religious and mystical quest reaching back to the origins of Western idealistic thought.

Yung Sik KIM, Seoul National University, Seoul, Korea

SOME PROBLEMS IN THE STUDY OF THE SCIENCE-RELIGION RELATIONSHIP IN TRADITIONAL CHINA

In the West, both science and religion were important elements in society, and it is natural that the two should have interacted in various ways. Moreover the objects of the Western science and religion, Nature and God, were inherently linked to each other in that the former was believed to have been created by the latter. Thus we see many instances in the Western history in which science and religion came to touch each other, generating significant influences in the development of each of them. But in China, neither science nor religion was very important in any period. Nor was there, in their ideas of science and religion, such an obvious link between the two. Study of the "science-religion" relationship in this situation has often tended to look for tenuous, or even imaginary, links and influences between the two elements that themselves were not important in their own context. For example, the role of Taoism for the Chinese scientific development has received an attention far beyond what it really deserves. The influence upon science of the "absence" of certain religious elements, also, has been among the more frequently discussed topics. These, of course, led to much interesting explanation of various particular ways Chinese scientific knowledge developed, or failed to develop. But they raise problems as well.

The present paper discusses mainly these problems. But it also raises other problems that come up when the "science-religion" relationship is studied in the context of traditional China. It indicates, among other things, that not only is the "science-religion" relationship culture-dependent but also is what we mean by "science-religion relationship".

Robert John Russell

Center for Theology and the Natural Sciences, GTU, Berkeley

THERMODYNAMIC IRREVERSIBILITY: CONTEMPORARY ARGUMENTS, HISTORICAL PERSPECTIVE, AND THEOLOGICAL SIGNIFICANCE.

In classical thermodynamics, closed systems evolve spontaneously from order to chaos. This process is describable by the increase of entropy and leads to a thermodynamic 'arrow of time.' Yet with statistical mechanics, entropy can be interpreted in terms of microscopically reversible processes, leaving the macroscopic arrow of time groundless. Now with contemporary non-equilibrium thermodynamics, the irreversibility of macroscopic processes seems irreducible. What is the significance of this shift for theological reflection in which a scientifically cogent view of nature plays a major role? This paper approaches this question by examining the history of irreversibility in thermodynamics over the past hundred years, noting its relation to theological concepts of order, chaos, time and history in Christian thought during this period.

Durant, John R.

Staff Tutor in Biological Sciences, Department for External Studie
University of Oxford, Engl

EVOLUTION AND ETHICS: ETHOLOGY, SOCIOBIOLOGY, AND THE
NATURALIZATION OF RELIGIOUS VALUES

The continually changing nature of science creates inevitable tensions in the relationship between knowledge and belief. The paper explores the implications of theoretical innovation for the attempt to articulate a scientifically informed theology through the example of mid-twentieth century debates about the relationship between evolution and ethics. In particular, it considers the consequences for liberal Protestant theology of the fundamental shift from classical ethological models of behavioural evolution in the 1950s and 1960s to "sociobiological" models in the 1970s and early 1980s.

Both classical ethology and sociobiology have offered theoretical explanations of "selfishness" and "altruism". What makes these explanations interesting for theology is not only that they claim to apply to human behaviour but also that they appear to have quite different moral implications. The paper compares and contrasts a number of attempts to integrate ethological and sociobiological theories into Christian theology. Analysis of the work of a number of biologists, philosophers, and theologians associated with the journal Zygon: Journal of Religion and Science forms the basis of the discussion, which identifies some of the constraints that are imposed upon both science and religion by the attempt to develop what R. W. Burhoe has termed a "scientific theology".

Viggo Mortensen

Lecturer, associate professor, Institute of Ethics and Philosophy of Religion,
University of Aarhus, Denmark

BEYOND RESTRICTION AND EXPANSION
A possible model for relating science and religion.

There is one model that has gained classical status in this century concerning the relationship between science and religion: The restrictionist model. Neoorthodoxy, existentialism and Wittgensteinian fideism all agree that science and religion belong to two separate and mutually exclusive realms of thought and experience; they neither interfere, nor do they conflict with one another.

The case against creationism has strengthened this standard model, but it has also been under severe attack, both from within the sciences themselves and within theology. As a result of such criticisms, an expansionist model has been put forward. (For the distinction between restrictionism and expansionism see L.R. Graham, Between science and values, 1981). A clear example of scientific expansionism is sociobiology; variations of expansionist arguments have been promulgated in the journal *Zygon*, and expecially by its founding editor, R.W. Burhoe (Toward a scientific theology, 1981).

In my paper I will defend the thesis that it is time to move beyond both restrictionist and expansionist models. Processthinking (process-philosophy and -theology) moves in this direction, but it is combined with a metaphysical system, i.e. Whiteheads, "which is no longer of currency even in an intellectual and philosophical climate deeply influenced by science." (A.R. Peacocke, Creation and the world of science, 1979).

The theoretical framework for my relating science and religion is taken from the Danish philosopher and theologian K.E. Løgstrup's (1905-1981) descriptive Metaphysics (I-IV, 1976-84), in which a phenomenological analysis open to scientific information and to religious interpretation forms the basis. The notion of complementarity is investigated to see if it can be used for a better determination of the relationship between the various descriptions and interpretations. The model has wide implications for theological anthropology. In this connexion, W. Pannenberg's thesis of the religious dimension of the "human" is of great interest. (Anthropologie in theologischer Perspektive, 1983). It also invites us to reevaluate the relationship between science and values, avoiding the extremes of value-free science and science-free values.

Wiebe E. BIJKER

Twente University of Technology, Enschede, The Netherlands

THE SOCIAL CONSTRUCTION OF BAKELITE
- towards a theory of invention

First, the integrated social constructivist approach to the study of science and technology will be summarized (cf. Pinch and Bijker, Social Studies of Science, 14 (1984), 399-442).
Second, the results of a case-study of the history of Bakelite will furnish illustrations of the concepts 'relevant social group', 'interpretative flexibility', 'technological frame' and 'inclusion'. For example, the social groups of Celluloid engineers and electro-chemical engineers have played very different but important roles. Associated with each social group is a technological frame which determines the problem solving strategy of that group. Baekeland could construct his plastic material because he had a low inclusion in the Celluloid frame and a high inclusion in the electro-chemical frame.
Third, it will be outlined how this set of concepts allows for a comparative analysis of different cases. This will be illustrated by applying them to studies such as Callon (1980), Constant (1980), Hughes (1983), Latour (1983) and Noble (1979), resulting in a proposal for what might constitute a theory of invention.

Edward W. Constant II
Associate Professor of History, Carnegie-Mellon University

INVENTION IN HIERARCHICAL PERSPECTIVE

Most discussions of technological invention focus on that which is patentable and take as that which is to be explained the presumed originality, creativity, or genius of the inventor. This primary concern with processes of variation has resulted in relative neglect of processes of selection in general and of systems context in particular. Following Donald Campbell and Herbert Simon, all complex technological systems comprise a nested hierarchy of subsystems, and in turn most often are nested within larger (sociotechnical) systems. Such systems are hierarchically decomposable in Simon's sense: "pieces" can be modified or substituted independently, subject only to interface constraints. Thus, given multiple level hierachy and decomposability, an immense variety of plausible inventions is possible.

These characteristics are illustrated by cable-tool and rotary methods of drilling oil wells. Both techiques are subsumed within a macro-system of oil exploration, production, transportation, refining, and distribution. The two methods of drilling operate on fundamentally different principles. Yet in practice, both share many components: derricks, pipe, prime movers, and so on. Many inventions at many levels are possible, only some of which would be patentable. That is, only a narrow middle range would be legal "inventions," others would be too low (or trivial or obvious), others too high, for example, the conception of a higher-level system which would not constitute a patentable device. Other examples of diversity, hierarchy, and decomposability include oil pumps and natural gas compressors, piston and turbojet aircraft engines, and electrical utility sytems.

Primary focus on the hierarchical nature of technological systems illuminates several long standing issues concerning invention: the difference between radical or revolutionary and incremental or developmental inventions, why most patented inventions are never used, and the cross-feed among technologies and systems. Moreover, this hierarchical perspective suggests that all inventive processes, technological, scientific, and perhaps even artistic, likely resolve into a common set of problem solving activities, and that there is little need for *ad hoc* explantations of creativity.

Reese V. Jenkins

Professor of History and
Director and Editor of the Thomas A. Edison Papers

EDISON AND THE ART OF INVENTION

Thomas A. Edison, as one of the most prolific inventors in history, introduced a large and diverse set of technical changes, including stock printers, quadruplex telegraphy, the carbon-button telephone transmitter, the phonograph, the incandescent lighting system, and the motion picture. All of these were of substantial commercial importance in the United States and Europe. Design was at the heart of his creative mechanical, electro-mechanical, and chemical inventions. The substantial number of his laboratory notes and sketches document the subject, composition, motions, and forms that constituted his creative designs.

Edison's creative work was largely focused on communications and production where his compositional techniques included continuous flow of materials, transmitting and receiving instruments that exploited the contextual knowledge of the receiving operator, and new means of storage and retrieval of information. Moreover, in developing a new composition (design), he often would start by working by analogy with earlier approaches that had been successful for him, but he would then often transcend the initial analogy in order to accomplish his innovative goals.

Some of Edison's compositional techniques reflect presuppositions, approaches, and ideas that were congruent with those of leaders in the fine arts, literature, music, and science of his era. Indeed, Edison's conceptual approach reflected the ideas of the late 19th century but, as with his artistic contemporaries, his creative work laid the foundation for the major cultural revolution that occurred early in the 20th century.

Prof. Peter Weingart, University of Bielefeld, Bielef
West Germany

FROM SOCIAL TECHNOLOGY TO TECHNOLOGICAL FIX- THE CONTROL OF PROCREATIVE BEHAVIOR

While eugenics and human genetics are seen in continui if perceived as the development of the scientific control of procreative behavior a distinction between the may be seen in a fundamental shift in their underlying technological philosophy. Eugenic schemes sought to implement hereditary theory by directing procreative beha vior directly and indirectly through social and politi measures ranging from changes of ethics, education, to marriage permits and sterilization. Thus they implied f reaching alterations of social values and institutions which proved to be insurmountable obstacles. This was monstrated by the implementation of racehygiene in Naz. Germany. --Due to advances both in medical technology and in genetics, techniques were developed which made massive state intervention characteristic of eugenic schemes obsolete. Genetic counseling, amniocentesis, selective abortion, in-vitro fertilization and possible future applications of recombination techniques are al applied within the framework of the medical profession are oriented to the individual rather than to the gene pool, and are applied on the priciple of voluntary and rational decision of the patient. The availability of these techniques has changed the values and institutior connected with procreative behavior drastically and cor tinues to do so. They do not require political coercior because the scientific premises on which they are based have been internalized which signifies an overall process of rationalization.

Dénes Nagy, Eötvös Loránd University, Budapest, Hungary

FROM τέχνη (TECHNĒ) AND 技 (JÌ) TO GENERAL TECHNICS (THE HISTORY OF TECHNOLOGICAL STYLES FROM THE ANTIQUITY TO THE RECENT TIME)

1. Historical survey (based on etymology)

τέχνη

- skill, cunning of hand (esp. in metal-working, also in ship's carpentering)
  e.g. Homer *Odyssea* 3.433, *Ilias* 3.61 (9th c. B.C.)
- craft, cunning (in bad sense)
  e.g. Homer *Odyssea* 4.455 (9th c. B.C.)
- way, manner, or means whereby a thing is gained
  e.g. Herodotus 1.112 (5th c. B.C.)
- an art, craft - a set of rules system or method of making or doing
  e.g. Plato *Phaedrus* 245a (5th-4th c. B.C.)
  Aristotle *Rhetorica* 1354a11 (4th c. B.C.)
              (see e.g. Liddel and Scott)
see e.g. Calepinus *Dictionarium* (1502, etc.)

技 Chin. jì
  Jap. gi
- skill, ability
- art, craft
  (see e.g. Nelson, Фельдман-Конрад - No 64.4)

Industrial revolution (18th c.)

- cunning by machine (<Gr. μηχανή = machine for lifting weights, crane, etc.)
- cunning by engine (<Lat. ingenium = talent, intellect, etc.; cf. genius)

handicraftsman
(homo faber) ↘ technician
                (homo technicus)
                                    engineer
Technique   Technical instruments   Technology
               Technical systems

                    ↓
               Technics
          技術 Jap. gijutcu

2. The Hungarian educational system of technics

Jonathan Liebenau

Business History Unit, London School of Economics, London, UK

A Case Study in the Transfer of A Medical Technology: The Production and Use of Diphtheria Antitoxin in Europe and the United States

The initial research on diphtheria antitoxin was done in Germany within academic institutions and hospitals. As soon as its feasibility was proclaimed, clinical trials were conducted in Berlin and Paris. The technical problems of scaling up production occupied medical researchers and industrial scientists through the 1890s, and American medical scientists visited the European pioneers even before the initial production problems were solved. From the mid-1890s the contrasting technical and medical goals, and the variety of institutional contexts, made a tremendous impact on the rate and direction of development. In Germany slow clinical progress was made, but there was rapid industrial development. In France the public health orientation of the Pasteur Institute narrowed the focus but insured a market for all the medication produced. The British context was quite different with little early interest by public health workers and only limited private initiative, leaving one commercial and one independent research laboratory with isolated development programs and no assured market.

Unguarded enthusiasm was the response in the United States. The transfer to the U.S. took place in commercial, municipal, and federal government institutions, and these hosts developed antitoxin unencumbered by the nagging problems some European producers still wished to solve. Though scientifically still behind German and French researchers, technically and commercially American producers were able to launch their antitoxin to the immediate benefit of public health campaigns and the emerging ethical medicines industry.

What characterises this transfer and diffusion is the initial simplification of the scientific complexities while concentrating on manufacturing and distributing the new product. Since the theory and practice associated with these novel products was unfamiliar to even sophisticated consumers, educating the market became a key element in the successful transfer. Production technologies improved to the extent that during the First World War U.S. producers were able to satisfy both the domestic market and provide vital supplies requested by the European Allies.

William F. Aspray

Associate Director, The Charles Babbage Institute

PATTERNS OF INTERNATIONAL DIFFUSION OF COMPUTER TECHNOLOGY, 1945-55

The paper examines the differing patterns of international diffusion of computer technology in its formative decade, 1945-55. Comparisons are made of technology flow between and out of the main producer countries (U.S., England, Germany); of flow from academic, government, and private industrial institutions; and of flow of competing technologies (analog and digital, electromechanical and electronic).

FRIDENSON Patrick

Maître-assistant, Université Paris X-Nanterre

THE TRANSFER OF AUTOMOTIVE TECHNOLOGY FROM FRANCE TO JAPAN IN THE

After World War II, Japanese automobile companies decided to improve their know how in automobile technology in order to enter the mass market and mass production. They signed thus licenzing and producing agreements of European models with leading European auto makers, under the close supervision (and with the cooperation of the State, represented by the MITI. The Hino Motors Company produced the 4 CV, a small car which was the largest sale of the biggest French auto firm, Renault. This production in Japan was not a great success in commercial terms. But it was a technological success. Production in Japan was not a mere replica of production in France. It offered several improvements. The Japanese engineer and workers learnt the recipes of mass production. Their French counterparts, who helped them closely, learnt the conditions and methods of technology transfer on less developed markets. It will be argued that this was one of the forgotten origins of the Japanese automobile boom and of the orientation of the French auto industry toward multinational production, as well as of the great careers of the initiators of this transfer.

Olav Wicken,   researcher

National Defence College Norway/Research Centre for Defence History,
Tollbugt. 10, 0152 Oslo, Norway

THE ROLE OF TECHNOLOGY TRANSFER IN THE ESTABLISHMENT OF A LOCAL
COMPUTER INDUSTRY

At the end of 1959, the 26 year old engineerer Yngvar Lundh
returned to the Norwegian Defence Research Establishment (NDRE)
after having spent 18 months at Massachusets Institute of
Technology, Lincoln Labratory. His stay in the US was of crucial
importance for the establishment of a Norwegian Computer industry.
However, more than seven years passed before the first computer
plants were put into operation based on Yngvar Lundh's basic ideas
from his stay at MIT.

When Lundh returned to NDRE, he managed to convince the
management of the research institute that digital technology was
of the utmost importance for future military - and civil -
industrial production. He became the leader of a group charged
with designing a small special purpose computer. This group soon
got the nick-name 'digital group', and the members started constructing a multi purpose computer for the ship-to-ship weapon
system 'Penguin' in 1962. In the years to come many of the
persons involved in the 'digital group' received scholarships to
spend a year at American academic institutions. Most of them
applied for and were admitted to MIT. In this way, close links
were established between the computer constructors at NDRE and the
scientists at MIT.

This was not the only division of NDRE which established
close ties with American academic research institutions. In the
1970's these connections formed an important link to advanced
American technology, as many of the NDRE researchers' colleagues
had moved into central positions in American companies, research
institutes or universities.

In 1967 members of the 'digital group' participated in the
establishment of two computer companies, both competing for a
military contract. Today, the company which lost this contract
is the dominating Norwegian computer company with growth rate of
about 40-50 per cent per year and a production value of about
NOK 1400 millions of 150 million US dollars at todays exchange
rate.

Paulo Bastos Tigre

Research fellow   Inst. de Ecomia Industrial

Transfer of Computer Technology to Brazil: A Short-cut to Self-Reliance?

Computer manufacturing has become one of the fastest growing business in Brazil with the help of favorable government policies. Sales of data processing equipment are expected to reach $2 billion in 1984 (from $30 million in 1978), more than half accruing to wholly Brazilian owned firms.

The government regulates imports and direct manufacturing by transnationals in order to save the most dynamic sectors of the computer industry for local manufacturers. These often combine foreign technology obtained through licensing agreements with their own product development. Technology suppliers include Europe an firms such as Nixsorf (West Germany) and Logabax (France) American peripheral equipment manufacturers, and Fujitsu from Japan.

Many local firms were able to develp the in-house capacity to adapt, extend and improve upon the imported technology. However they face several difficulties. These include the problem of maintaining techological development with limited resources while facing direct or indirect competition with large foreign multinationals. The 32-bit "supermini" recently launched by leading American and Japanese mini-makers is a new challenge for local firms. Most Brazilian minicomputer manufacturers claim that they have technical ability to develop a 32 bit supermini in-house. But the development of such advanced equipment could be too risky, time consuming, and expensive for local firms. In addition there is some scepticism about the competitive strength of a locally designed supermini in a market where standards had been set by foreign technology. Alternatively local manufactures are now shopping around for new technology transfer agreements.

The aim of the paper is to examine the reasons for acquiring technology through licensing, and their implications for competitive industry structure and technological self-reliance. Licensing will be compared with existing alternatives for obtaining technical know-how, including local product development and local manufacturing by subsidiaries of multinational firm. The paper is based on information gathered by the author in 1980 for PhD Thesis (Completed in 1982). This will be cross-checked with a new round of interviews with executives of Brazilian computer firms involved in the licensing business.

This will enable drawing new conclusions on the opportunities, advantages and disadvantages of the technology transfer process.

Eda Kranakis

Yale University, New Haven, Connecticut, USA

A COMPARISON OF TECHNOLOGICAL STYLES IN FRANCE AND THE U.S., 19TH C.

This paper contrasts the national technological traditions of France and America during the first half of the 19th century by comparing the evolution of suspension bridge technology in the two countries. The contributions of three individuals in particular are examined: James Finley, a Pennsylvania judge who was among the first to design a level-roadway suspension bridge; Claude Louis Navier, an engineer with the French Corps des Ponts et Chaussees, who published the first comprehensive theory of the suspension bridge and who undertook to build a grand suspension bridge in Paris; and Marc Seguin, a French entrepreneur who designed a large number of wire-cable suspension bridges.

Americans evolved a technological tradition that emphasized functionalism in design and an inductive, experimental research methodology. Finley's work represents an early example of this trend. Finley carried out a series of experiments from which he induced some important "laws" about suspension bridges. This knowledge enabled him to work out a simple, practical method to proportion suspension bridges.

France exhibited conflicting technological traditions: one associated with the state engineering corps and the "Grandes Ecoles"; and one associated with the emerging community of engineers in the private sector. State engineers like Navier favored deductive engineering theory over empirical research. Navier developed an elegant and complex mathematical model of the suspension bridge from which he deduced a number of design principles as well as new knowledge about the behavior of this type of structure. French engineers in the private sector, like Marc Seguin, however, rejected the state engineers' theoretical approach; they favored practical tests and experiments. For example, Marc Seguin carried out numerous tests to prove the superiority of wire over chain cables, an issue that Navier simply ignored. However, since state engineers had considerably more power and prestige than engineers in the private sector, their theoretical approach tended to predominate in France during the 19th century. This fact helps to account for the failure of French engineers to develop a strong industrial research tradition such as evolved in the United States.

Svante Lindqvist

Royal Institute of Technology, Stockholm, Sweden

"INTO THE JAWS OF THE UTILITARIAN BEAST" - SWEDISH VIEWS OF AMERICAN TECHNOLOGY IN THE 19TH CENTURY

During the 19th century, Swedish engineers' opinions of the USA as an industrial nation changed. There was a ri in esteem which is not an absolute measure of the advance of the USA, but rather a relative measure of th differences between the two countries.

Until about 1860, Swedish engineers regarded Sweden ahe both quantitatively and qualitatively. During the perio 1860-1890, it was noted that the USA was producing larger quantities than Sweden, but Swedish technology w still considered of superior quality. After about 1890, however, it became necessary to accept that the USA was also ahead of Sweden in quality.

During the period when Swedish engineers acknowledged only the USA's <u>quantitative</u> superiority, they criticize the preoccupation with output. The USA had gone too far they said, in its greed for profits. The engineer and author Per Hallström, who worked at a factory in Philadelphia in 1889-90, even likened his stay in the USA to "running straight into the jaws of the beast of utility". But when Swedish engineers found themselves also obliged to recognize the <u>qualitative</u> superiority o the USA, they began instead to direct their criticism a their own country and its "antiquated and inadequate methods".

The changing view among Swedish engineers of the USA as an industrial nation illustrates the different national styles of technology in the two countries; differences that had their roots in differing cultural values.

Pamela E. Mack

Assistant Professor, Clemson University

"CONFLICTING STYLES IN AN INSTITUTIONAL CONTEXT:
SCIENCE, TECHNOLOGY, AND THE SPACE PROGRAM"

      Historians of the U.S. space program have identified the conflict between scientists and engineers as a key theme. Scientists form an unusually cohesive and well-differentiated community, and their style was often in conflict with the style of the space agency. This paper will use examples of the interaction between scientists and the National Aviation and Space Administration to examine a more general pattern: the relationship between the institutions at which people are employed and the professional communities to which most engineers and scientists owe some loyalty.
      The approach to analyzing the interaction of these communities will be to use a matrix as a basic analogy. In the simplest case people belong both to a column of the matrix--a project team working on a particular project within a bureaucratic chain of command--and a row--a professional community in a particular scientific or engineering specialty. The relative importance of these two communities in a particular situation is closely interrelated with the process and character of technological change. This analogy is more useful than most models currently used by historians of technology for examining technological change in large research and development organizations because it provides a way of examining the link between institutions and scientific and technological knowledge and practice.

Dr. John M. Staudenmaier, S.J.

Assistant Professor: History of Technology, The University of Detroit

TECHNOLOGY, USA STYLE: 20TH CENTURY IMBALANCES

One major value dichotomy is embedded in the design of those technologies which have come to be "normal" in 20th century United States. "Normal" here means both "ordinary" and "normative" and refers to technologies such as electricity, electronic communication systems (telephone, radio, television, computer), the automobile, mass production and consumer marketing systems, etc. The dichotomy, "SYSTEMIC STANDARDIZATION REPLACES NEGOTIATION" achieved hegemony in the United States about 1880 and has dominated U.S. technological style through this century. At its core, the style represents a strong preference for solving problems through increasingly sophisticated centralized systems which bypass the messy uncertainties of negotiation with exogenous contextual factors by drawing them into the larger system and transforming them into conforming functional components.

The U.S. style emerged from and strongly reinforces that hatred of vulnerability and uncertainty that marks all processes of negotiation among mutually exogenous actors, whether personal, institutional, or ecological. Systemic standardization is based on the assumption that the technological practicioner can focus nearly all attention on the complex problems of systemic design and maintenance leaving the nonsystemic contextual environment to take care of itself. A century's commitment to this style has led to an "imbalanced" national culture, one where precise elegance in system design has flourished and the subtleties of negotiation has atrophied.

The author will argue that no national culture can ignore the processes of negotiation forever and that the cumulative effects of the U.S. style constitute a crisis of technological style calling for a renewal of the ability to negotiate on all levels of national praxis.

Hans-Werner Schütt

Professor   Technische Universität Berlin, West Germany

ZUR GESCHICHTE DER MODERNEN CHEMISCHEN NOMENKLATUR

Mit finanzieller Unterstützung der Stiftung Volkswagenwerk wurde ein Progamm zur Geschichte der modernen chemischen Nomenklatur seit Gründung der IUPAC im Jahre 1919 durchgeführt, dessen Ergebnisse allerdings noch nicht völlig ausgewertet sind. Die Untersuchung stützt sich auf Material aus den Archiven der International Union of Pure and Applied Chemistry in Oxford, England, und des Verlags der Chemical Abstracts in Columbus, Ohio. Im Mittelpunkt der Untersuchung steht die Geschichte der IUPAC-Kommission für anorganische Chemie, ihre Zusammensetzung, ihre Arbeit und die Methoden, mit denen sie ihre Nomenklaturvorschläge in die multinationale Gemeinschaft der Chemiker einführt.

Michel Callon

Professeur, Ecole des Mines de Paris, Centre de Sociologie

de l'Innovation

ACTOR-NETWORKS AND THE ELECTRIC VEHICLE: THE STUDY OF TECHNOLOGY AS A TOOL FOR SOCIOLOGICAL ANALYSIS

Generally speaking, the history and sociology of technology emphasizes the relationships and influences linking artefacts to the social contexts in which they are developed. Certain writers underline the autonomy of technical systems or networks; others demonstrate the influence of social, economic and cultural factors. This paper gives a different viewpoint. Taking the electric vehicle as an example (1960-1975), it tries to show the possibility of using the study of technological development as a tool for sociological analysis.

Some studies (Hughes, Jenkins) demonstrate that engineers and scientists who develop new technologies constantly raise hypotheses regarding social structures: identity or behaviour of social groups, evolution of needs and values. Translated into technical decisions, these suppositions on the composition of surrounding society shape the artefact. The success or failure of the artefact, its transformations are consequently tests of the (relative) veracity of the hypotheses. To follow the development of an artefact is thus to explore society with the engineers and researchers who, through their projects and achievements, learn to understand it and to whose development they contribute. The study of technological development could thus be used as an instrument of sociological analysis.

Starting with a criticism of the notions of system and network, this paper presents the concept of the actor-network which allows us to understand how engineers simultaneously explore society and define technological artefacts which are adapted to, or which can transform, it.

This notion emphasizes the heterogeneity of elements (cultural, technical, social or economic) which are linked by the actors when they define and disseminate a technical device. The actor network can be distinguished from a simple network because its elements are both heterogeneous and mutually defined in the course of their association. It can be distinguished from a simple actor by its structure which is an arrangement of heterogeneous and remote entities.

Sylvia Doughty Fries

Chief Historian, National Aeronautics & Space Administration (NASA)

The U.S. Space Station as Technological System

As far back as the mid-nineteenth century dreamers and designers of space travel and exploration have envisioned, among their fanciful and not so fanciful schemes, the presence of inhabited orbiting outposts, or "space stations." Within NASA, the U.S. space agency, the concept of a space station has had one of the longest design histories of any item in the agency's repertoire of space vehicles and satellites.

This paper will highlight the quarter-century evolution of NASA's space station designs by describing three designs from three successive periods (pre-Apollo, post-Apollo, and post-Shuttle). In so doing, it will focus on each space station as a system and attempt to demonstrate how environmental and technological constraints determined the systems characteristics of each successive design.

Thomas P. Hughes, Professor of the History of Science and Technology
University of Pennsylvania

REVERSE SALIENTS, CRITICAL PROBLEMS, AND EVOLVING SYSTEMS

The concepts of reverse salients and critical problems help explain the dynamism of evolving systems. The author has used these concepts in a published worked entitled NETWORKS OF POWER, but the definitions and the case histories pertain particularly to electric light and power systems. To suggest the general applicability of the concepts, this paper provides definitions and examples taken from the broad sweep of the history of technology and applied science. Sources used are both secondary and primary.

John Law

Department of Sociology and Social Anthropology, University of Keele

VESSELS AND THEIR CONTEXT: THE PORTUGUESE MARITIME EMPIRE

In the 1480s the Portuguese created a method of astronomical navigation which, when combined with the fifteenth century Iberian development of the multimasted mixed rigged vessel, made it possible to exercise control over a large part of the Indian Ocean. This paper describes elements of the navigational and maritime system developed by the Portuguese as they sought to control political and economic events in India. In particular, it describes the work of the Commission established in 1484 by King John II to resolve the navigational problems that resulted when vessels sailed beyond European waters.

It is suggested that these innovations in marine technology and navigation should not be seen in isolation, but should rather be interpreted as forming part of a larger system where a range of elements (artefacts, people, inscriptions, economic forces etc) were assembled together. Such "heterogeneous engineering" which is a characteristic of system-building, involves (amongst other features) the construction of (a) artefacts and (b) contexts or "envelopes" for those artefacts that allow these to retain their integrity and coherence. In this paper the relationship between artefacts and envelopes is considered for both Portuguese vessels and navigational instruments. In particular, it is noted that the social, geographical and natural envelopes of these artefacts were enhanced in a process in which such artefacts formed a part of each others' envelopes. Non-artefactual features of these envelopes are also considered, as is the extent to which the same approach may be usefully applied to the other components (notably people and texts) that made up the Portuguese maritime system.

Donald MacKenzie

Lecturer in Sociology, University of Edinburgh

THE MISSILE ACCURACY SYSTEM

A major technological development of the last 25 years has been the growth in strategic missile accuracy, with all the implications that has for the security of land-based 'deterrent' systems and the possibility of a nuclear 'first strike'.

This paper is a preliminary attempt to understand that development. It applies to the missile accuracy system the model of technical change proposed by Thomas Hughes in <u>Networks of Power</u>, showing both strong similarities and significant differences to the case discussed by Hughes. The paper shows the genuinely <u>systemic</u> nature of the technology of missile accuracy, underlining how areas of apparently 'pure' science become militarily crucial through their place in the system. The distinctive dynamic of technical change of the system is identified, and its relations to organisational interests and wider military and political factors are discussed.

The system of technologies whose outcome is missile accuracy includes inertial instrument design, computer technology and microelectronic miniaturisation, geodesy and mapping, satellite surveillance, geophysics, hydrodynamics and materials science. Technical change in this overall system, and in each component subsystem, has focused in a goal-oriented fashion on the elimination of what Hughes calls 'reverse salients' - here, perceived barriers to the growth of missile accuracy. The nature of these reverse salients is not simply 'given'. Participants dispute what the true reverse salients are, and major organisational interests bear upon the outcomes of these disputes. Particular organisations - for example the crucial Charles Stark Draper Laboratory in Cambridge, Mass. - have over the years developed major investments (in terms of personnel and skills as well as finance) in particular directions of technical advance; these contribute to distinctive patterns of technological 'momentum'. But these patterns are not self-explaining, or to be explained simply by the preferences of organisations such as the Draper Lab. They interact with national circumstances (eg possible differences in technological style between the US and USSR in these matters), with the situations of 'consumer' organisations (such as the US Navy and Air Force), and with the long-standing and deeply entrenched preference, in both the US and USSR, for a 'war-fighting' rather than simply 'deterrent' approach to nuclear targeting.

The missile accuracy system has, despite its entrenchment, achieved nothing like the degree of control over its social environment that the electricity supply systems discussed by Hughes have. Both the missile accuracy system as a whole, and particular institutions such as the Draper Laboratory, remain vulnerable to external change

Jean Claude Martzloff

Centre National de la Recherche Scientifique (Paris) et Université de Tokyo

L'APPRENTISSAGE DE LA GEOMETRIE PAR L'EMPEREUR KANGXI VERS 1690-1692 D'APRES THE JOURNAL DE JOACHIM BOUVET

S'il est bien connu que les six premiers livres des <u>Eléments</u> d'<u>Euclide</u> ont été traduits en chinois par Matteo Ricci (1552-1610) et Xu Guangqi (1562-1633) au début du XVIIe siècle et que la version chinoise complète du même ouvrage n'a été achevée que dans la seconde moitié du XIXe siècle grâce aux efforts d'A. Wylie (1815-1887) et de Li Shanlan (1811-1882), en revanche, il est moins bien connu que plusieurs versions, chinoises et mandchoues des mêmes <u>Eléments</u> ont aussi été effectuées à la fin du XVIIe et au début du XVIIIe siècle. Certes, il est vrai que ces dernières diffèrent complètement de la version de Ricci/Xu, tant du point de vue de la rigueur des démonstrations que de celui de l'agencement des théorèmes, mais, comme elles portent toutes le même titre que l'ouvrage initial (chinois: jihe yuanben; mandchou: <u>gihe yuwan ben</u>, c'est-à-dire, littéralement "Livre élémentaire relatif à la mesure des grandeurs" et non pas "Eléments de géométrie" comme on l'écrit souvent), il en est résulté des confusions. Aussi, nous nous proposerons de montrer dans la présente communication que:
1- La source européenne principale de toutes ces versions (qui sont toutes manuscriptes, à l'exception de celle qui figure dans l'important <u>Shuli jingyun</u> (ca 1723), est un manuel d'enseignement d'un jésuite français, le Père Gaston Ignace Pardies (1636-1673). Ce manuel (Eléments de géométrie, première éd., Paris 1671) ne contient pas seulement que des résultats figurant dans Euclide, mais aussi d'autres apparaissant dans Archimède et Appolonius.
2- L'Empereur Kangxi a appris la géométrie élémentaire en se servant d'adaptations chinoises et mandchoues d'un tel manuel, et non pas à partir de l'ouvrage de Ricci/Xu.

Le premier historian qui a proposé 1- est le Père Henri Bernard Maître mais, comme récemment des historiens chinois ou japonais en ont contesté la validité, nous examinerons à nouveau la question.

La validité de 2- n'est pas généralement reconnue. Par exemple, Wang Ping (Academia Sinica, Taipeh) pense le contraire. En fait, jusqu'à présent, les seuls renseignements contenus dans le célèbre ouvrage de Joachim Bouvet (1656-1730), <u>Portrait historique de l'Empereur de la Chine présenté au Roy</u> (Paris, 1697), ne permettaient pas de se faire une opinion précise. Par contre, le journal de J. Bouvet (Bibliothèque Nationale, Paris, Mss fr. 17240) contient une description détaillée, jour par jour et théorème par théorème, de l'apprentissage de la géométrie élémentaire par l'empereur Kangxi vers 1690.

Catherine JAMI

Ecole Normale Supérieure de Jeunes Filles, France

AN EXAMPLE OF THE JESUITS' INFLUENCE ON CHINESE MATHEMATICAL ACTIVITY IN THE 18TH CENTURY

Since the beginning of the 17th century, the knowledge brought to China by the Jesuit missionnaries was the basis of a revival of Chinese science. In the field of mathematics, the Chinese not only assimilated Western knowledge, but also developed an independent activity and rediscovered their own tradition.

Ming Antu's mathematical work (which can be situated towards the middle of the 18th century) is a good illustration of this situation : he demonstrated some formulae of development of trigonometrical functions in power series ; some of these formulae had been introduced into China by a French Jesuit missionnary, P. Jartoux.

The analysis of Ming Antu's book, the Ge Yuan Mi Lü Jie Fa "Quick Method for Determining Precisely the Ratios of the Circle", puts into light the specific modalities of the meeting of two mathematical traditions.

As for the sources of Ming Antu's work, his subject and methods (some of which belong to the field of euclidian geometry) refer mostly to Western mathematics. On the other hand, the most recent developments of calculus had not reached China at his time

The mathematical language he uses comes partly from Chinese tradition, partly from translations of western terminology.

Some idea of the Chinese's point of view concerning the Western influence on their science can be drawn from the comments and the forewords of the book : they considered the new development of mathematics at Ming Antu's time, including the works of the Jesuits themselves as the continuation of their ancient tradition.

Lenore Feigenbaum

Assistant Professor, Northeastern University, Boston, USA

THE TAYLOR-MONMORT CORRESPONDENCE AND THE MATHEMATICAL COMMUNITY OF THE EARLY 18TH CENTURY

If the development and promotion of the calculus was a collective European effort, so too was the infamous priority dispute surrounding its consolidation in the hands of Newton and Leibniz. Embroiling the entire mathematical community, the conflict fostered alliances and antagonisms along national lines and swept neutral figures into the fray as confidants and intermediaries. Although much is known about the principal characters, comparatively little has been published about the minor figures who comprised the growing but fragmented community. Competent mathematicians in their own right and highly influential in the progress and diffusion of the new calculus, these less prominent figures contributed to, and at the same time were victimized by, the climate of prejudice, hatred, and deceit engendered by the quarrel and by the secondary disputes that erupted in its wake.

The correspondence between the neutral Frenchman Pierre Remond de Monmort (1678-1719) and the English Newtonian Brook Taylor (1685-1731) during the years 1715-1719 provides us with a valuable perspective on those who, if not centrally involved, were nevertheless deeply affected by the controversy. As an intermediary between John Keill and Taylor on the one hand and the Bernoullis on the other, Monmort informs us of Taylor's activities and colleagues, including his uneasy relationship with Newton, and about the inner workings of the circle around Leibniz' collaborator and successor, Johann Bernoulli. Since Monmort's own role is controversial, having earned him the reputation of troublemaker as well as peacemaker, we learn first-hand about some of the psychological, social, and mathematical repercussions of the priority battle on the wider community. As a primary actor in his own dispute with De Moivre and as an important figure in the French scientific community, Monmort paints a vivid picture of French mathematics in the early 18th century and helps illuminate the complex personal and scientific connections between mathematicians on both sides of the English Channel.

Jeremy Gray

Faculty of Mathematics, Open University, Milton Keynes, England

Mathematical Prize Competitions of the late 19th Century

Throughout the 19th century many of the learned societies of Europe ran prize competitions or awarded medals in mathematics. Centres included Paris, Berlin, Leipzig, Brussels, London and Rome. A study of these competitions provides information about the contemporary definition of mathematics and the relative importance of different topics in mathematics, as well as the composition of elites (panels and prizewinners). Accordingly it contributes to the attempts to build up a picture of the various scientific communities of the period capable of illuminating the connection between the perceived aims of science and its social context. In my talk I shall concentrate on the implications for mathematics, specifically geometry, and discuss what can be inferred about the importance of the subject at the research and teaching levels.

Pierre DUGAC

Directeur d'Etudes, Ecole Pratique des Hautes Etudes, Mathématiques

LE RAYONNEMENT SCIENTIFIQUE D'HENRI POINCARE A LA LUMIERE DE SA CORRESPONDANCE

La vaste correspondance d'Henri Poincaré, dont les photocopies se trouvent actuellement au Séminaire d'Histoire des Mathématiques à l'Institut Henri Poincaré, a été rassemblée autour du noyau constitué par les lettres qui appartiennent aujourd'hui à son petit-fils François Poincaré. Ces lettres ont été déjà utilisées, en partie, dans le livre que A. Bellivier a consacré, en 1956, à Henri Poincaré. Elles ont été ensuite retrouvées par A.I. Miller qui en a établi une liste, ainsi que des microfilms. Enfin, G. Masotto a repris contact avec F. Poincaré et il nous a permis de profiter du travail accompli par A.I. Miller.

Cette correspondance montre à l'évidence l'extraordinaire rayonnement scientifique du génie de Poincaré, ainsi que son activité exceptionnellement féconde dans de nombreuses sciences aussi bien théoriques qu'appliquées (elle est en particulier très révélatrice en ce qui concerne l'intérêt de Poincaré pour les expériences de la physique). Avant de nous tourner vers les sciences mathématiques, qui formeront l'essentiel de notre propos, on peut signaler quelques correspondants de Poincaré dans d'autres domaines. On peut citer ainsi les physiciens H. Becquerel, P. Curie, P. Duhem, H. Herz, H.A. Lorentz, A. Michelson et W. Thomson, le chimiste H. Le Chatelier et les géologues M. Bertrand et A. Dubrée.

Il n'est pas question de donner ici la liste, même abrégée, de la pléiade des mathématiciens qui ont correspondu avec Poincaré et qui ont dominé la mathématique entre 1880 et 1910. D'ailleurs, le fascicule 7(1986) des *Cahiers du Séminaire d'Histoire des Mathématiques* publiera des informations et des lettres de la *Correspondance d'Henri Poincaré avec des mathématiciens*. Cette correspondance permet de comprendre comment se sont développées certaines des idées dont l'influence fut décisive dans l'évolution des mathématiques. Elle nous renseigne également sur l'oeuvre et l'influence de Poincaré.

Indiquons rapidement quelques points, entre parenthèses, concernant certains des mathématiciens : P. Appell (sur ses recherches de 1881 voisines de celles de Poincaré), G. Darboux (les réticences du rapporteur de la thèse de Poincaré sur son manque de clarté), G. Halphen (sur l'inspiration, puisée chez Poincaré, de ses recherches sur les équations différentielles), F. Klein (sur l'origine des nouvelles fonctions introduites par Poincaré et l'importance que Klein attachait, encore en 1902, à l'oeuvre de Poincaré) et G. Mittag-Leffler (sur le génie d'invention de Poincaré qui arrive à surmonter des difficultés mathématiques qui allaient tourner presque à une "tragédie", à propos d'un prix qui lui était attribué, lorsqu'on y découvre une grave erreur).

Jean CASSINET
Maître-Assistant. Docteur.
Université Paul Sabatier .TOULOUSE . France

UN INEDIT DE ROBERVAL(1645) SUR LE LIVR V DES ELEMENTS D'EUCLIDE ET SA TRANSMISSION VERS L'ITALIE.

Le registre de manuscrits n°1531 de la Bibliothèque Municipale de Toulouse(France),contient 11 copies de textes de FERMAT effectuées en 1644-45,pour le compte du mathématicien romain Michelangelo RICCI,au cours des voyages en Italie de DU VERDUS et du père MERSENNE.((1) & (2)).

Ce registre contient un texte inédit de G.Personne de ROBEF VAL de 39 pages(3) développant en 22 Théorèmes la théorie de l'ordre entre rapports de grandeurs,à partir de la définition n°7 du Livre V des Eléments d'Euclide(4).ROBERVAL démontre qu la relation ainsi définie est totale(appendix à la définition complétée par lui)asymétrique(Th.2)transitive(Th.5).Il mon tre la compatibilité de la relation avec : la multiplication des rapports par les rationnels(Th.6)la multiplication de 2 rapports "simplifiables"(Th.16,17)l'addition de 2 rapports de même dénominateur(Th.19,20);il montre le lien entre ordre et opérations habituelles sur les rapports(inversion,alternative,composition,division)ce qui est la partie non originale du texte(Th.10,11,12,13,14,18);une partie très originale concerne la "densité" du demi-groupe ordonné des rapports de grandeurs (Th.7,8,9,21) ce dernier traitant de grandeurs "non-numériques" et démontré par la méthode "d'exhaustion"
$((\forall H)(\forall K)((A+H)/B > C/D \ \& \ (A-K)/B < C/D \implies A/B = C/D)$
Enfin le Th.15 est la P.18 du Livre V ; il en donne deux démonstrations(une directe l'autre non) débarrassées de l'utilisation du postulat d'existence de la $4^{è}$proportionnelle à 3 grandeurs données.

C'est un texte remarquable,jamais égalé sur ce sujet.

(1).Paintandre R.(Mém.Acad.Sc.Toulouse(14)129(1967)77-87)
(2).Costabel P (Revue d'Hist.des Sc(1969)n°2.155-162)
(3) Texte en cours de publication avec commentaires par MM. J.Cassinet(Toulouse-France)& K.Hara(Osaka-Japon).(4)Heiber

Hans Niels Jahnke

University of Bielefeld, F.R. of Germany

SIMILARITIES AND DIFFERENCES OF BRITISH AND GERMAN REACTIONS TO FRENCH MATHEMATICS IN THE EARLY 19th CENTURY

At the beginning of the 19th century the then leading French mathematics was broadly received in Germany and Great Britain. This reception showed a number of similarities which were most clearly revealed in the analogous foundations of algebra and analysis developed by M. Ohm (1792-1872) and G. Peacock (1791-1858).

Historians of mathematics consider these systems predominently as a step towards an abstract view of algebra. But on the background of the contemporary context these systems may equally well be interpreted as a "compromise" of analytical and synthetical methods or as an attempt to combine the freedom of analytical procedures with a synthetical bottom up-approach. The viewpoint of compromise was present in various forms with many other German and British authors, too (for instance Grassmann, Steiner, DeMorgan, Hamilton, Whewell).

This taking into account of synthetical viewpoints seems to reflect the specific cultural, philosophical and institutional conditions which the analytical method encountered when transmitted to Germany and Great Britain.

The paper tries to explore this idea by comparing the two countries under mathematical, philosophical and institutional viewpoints.

Helena M. Pycior

Associate Professor, History, University of Wisconsin-Milwaukee

TRANSATLANTIC LINKS: BRITISH-AMERICAN MATHEMATICAL TIES OF THE
NINETEENTH CENTURY

   This paper will begin to assess the influence of British
mathematics and mathematicians on the development of mathematics
in the United States of the nineteenth century. It will describe
and analyze modes of communication between British and American
mathematicians; common use of mathematical textbooks;
transatlantic mathematical honors, such as honorary memberships
bestowed on American mathematicians by British mathematical and
scientific societies; visits across the Atlantic; and similar
styles of mathematical research and writing, with primary examples
taken from algebra.

Luboš Nový

Department of the History of
Science and Technology, Institute of Czechoslovak and
General History of CSAS

THE DEVELOPMENT OF CZECH MATHEMATICS: NATIONAL AND SCIENTIFIC IMPLICATIONS

The propagation of mathematical knowledge is encountering a number of obstacles even in the modern era. These obstacles take a concrete historical form which is determined by, among other factors, the way in which mathematics is incorporated in the life of a particular social unit. The purpose of this paper is to point out the complexity of this process using the development of Czech mathematics in the 2nd half of the 19th century, which may serve as a certain model with more general validity, as an example.

In the 2nd half of the 19th century, the number of mathematical journals being published increased. Their network was becoming increasingly denser particularly in Europe. The number of published papers increased to such an extent that a reference journal began to be issued in 1868. An apparently uniform system of mathematical information flow was thus created. On the other hand, the national movement was also developing and endeavouring to achieve general progress in the national culture as a whole, including science. In the Czech Lands, this process had a marked effect on the development of mathematics in the 2nd half of the 19th century. A large effort was also being made to establish Czech as a "scientific" language; concurrently, the number of Czech specialists, originating from the unilingual Czech milieu, also increased. These were the two reasons for wanting to publish original results in Czech. This in turn influenced the publication efforts of mathematicians like V.Šimerka, V.Skuherský, F.Studnička and Eduard Weyr; the Czech Mathematical Society was founded (1862) and began to publish a journal in Czech. The Czech mathematicians are cognisant of foreign results, but their work is little known abroad. F.Studnička also supplied a large amount of information about Czech results for the reference journal. This process ahd unfavourable consequences for the appreciation of Czech achievements at home and abroad.

Of course, the Czech mathematicians were able to write comprehendingly for other Slavonic nations. That is why they earned merit for the development of mathematics in Yugoslavia (e.g. Zahradník) and Bulgaria.

The only way of solving the outlined situation as and is to publish parallelly in Czech (or Slovak) and in one of the world languages. These efforts were already in evidence in the Czech Lands towards the end of the 19th century, however, they had their difficulties which took a long time to overcome.

Erwin Neuenschwander

Universität Zürich, Zürich, Switzerland

MATHEMATISCHE ZEITSCHRIFTEN IM 19. JAHRHUNDERT UND IHR BEITRAG ZUM WISSENSCHAFTLICHEN AUSTAUSCH ZWISCHEN FRANKREICH UND DEUTSCHLAND

Die Bedeutung der mathematischen Fachzeitschriften für den wissenschaftlichen Austausch innerhalb der heutigen mathematischen Gemeinschaft ist unbestritten. Trotzdem existiert bis jetzt keine Gesamtstudie über die Entwicklung der mathematischen Zeitschriften und die bisher vorhandenen Detailstudien zu diesem Thema sind relativ selten und meist auf eine einzige Zeitschrift beschränkt. Diesem Sachverhalt versucht ein von uns in den Jahren 1980-1984 durchgeführtes, von der Stiftung Volkswagenwerk und der französischen Regierung unterstütztes Forschungsprogramm Rechnung zu tragen. Es beruht auf der seitenweise Durchsicht von vier exemplarisch ausgewählten mathematischen Zeitschriften (Journale von Crelle und Liouville, Archiv von Grunert, Nouvelles Annales) für die Jahre 1830-1880 und der Auswertung verschiedener Nachlässe, Briefwechsel und Verlagsarchive (Nachlass von Liouville, Briefwechsel von Borchardt, Briefwechsel zwischen Hoüel und Darboux usw.). Die hierbei aufgefundenen Resultate und Dokumente wurden einstweilen in der Form eines Preprints publiziert.

Bezüglich der im 19. Jahrhundert üblichen Usanzen und Editionsprizipien stellten wir fest, dass die Herausgeber bei der Annahme von Artikeln häufig uneinheitlich vorgingen. Während sie bei ihnen bekannten Autoren zum Teil Arbeiten akzeptierten, ohne diese überhaupt je gesehen zu haben, zogen sie in anderen Fällen zur Beurteilung auch befreundete Mathematiker hinzu. Zurückgewiesene Arbeiten wurden von den Autoren häufig bei anderen, meist zweitklassigen Zeitschriften eingereicht. Grundsätzlich zeigte sich, dass die meisten damaligen mathematischen Zeitschriften zunächst einmal Publikationsorgane für Autoren aus dem eigenen Lande waren. Zur Ergänzung nahmen viele Zeitschriften Uebersetzungen, Zweitabdrucke oder Bearbeitungen von bereits anderswo erschienenen wichtigen Arbeiten auf, womit sie einen entscheidenden Beitrag zum internationalen mathematischen Austausch leisteten.

Die systematische Durchsicht mathematischer Zeitschriften bietet ferner ein bisher kaum genutztes Mittel, die interne mathematische Entwicklung und die Einwirkung externer Faktoren im Detail zu studieren, wie anhand mehrerer Fallstudien demonstriert wurde (Aufnahme der Schriften von Cauchy in Deutschland, Ausbreitung der Theorien von Riemann und Weierstrass, Wechselwirkungen zwischen dem Aufschwung der Mathematik und der politischen Einigung in Italien).

Literatur: E. Neuenschwander, Die Edition mathematischer Zeitschriften im 19. Jahrhundert und ihr Beitrag zum wissenschaftlichen Austausch zwischen Frankreich und Deutschland. Preprint: Mathematisches Institut der Universität Göttingen, April 1984.

AUJAC Germaine

Professeur émérite à l'Université de TOULOUSE-LE MIRAIL

La transmission du texte d'Euclide : les définitions du livre V

Dans le livre V d'Euclide, la définition 17 (éd. Heiberg) se compose de deux formules présentées en alternative mais qui ne sont pas équivalentes. On retient généralement pour la définition du di'isou logos la première formule, que je crois le résultat d'une interpolation : elle énonce dans un libellé étranger aux habitudes euclidiennes le théorème qui sera démontré en V, 22, ce qui oblige à taxer Euclide d'incohérence quand il utilise le terme sans appliquer ce théorème (cf. V,3 et note de Heiberg). La définition authentique du di'isou logos me paraît être la seconde formule, semblable par son libellé aux définitions précédentes, et qui rend compte de tous les emplois du terme non seulement chez Euclide mais chez les géomètres postérieurs. L'erreur s'est d'ailleurs introduite dans le texte à date très ancienne puisqu'elle se trouve dans tous les manuscrits, théoniens ou préthéoniens. Elle est probablement le résultat d'une contamination avec la définition de la proportion réglée, que livrent à la suite certains manuscrits de la tradition théonienne, et que je crois nécessaire de rétablir dans le texte avant la définition, symétrique, de la proportion déréglée (déf. 18 chez Heiberg).

John Stroyls

Dhahran, Saudi Arabia

THE MIFTĀH AL-ḤISAB AND THE SHU-SHU CHIU-CHIANG:
A CIRCLE OF IDEAS

    There is much speculation about possible connections between Medieval Near Eastern and Chinese work on the numerical solution of polynomial equations in one variable. This paper examines this speculation with reference to: (1) cultural connections between the Near East and China (especially during the Mongol period), (2) transformations of mathematical methods required by different computational aids (abacus, dust-board, paper-pen), (3) extant texts, of which the Miftāh and Shu-shu are but two of many considered.

R.C. Gupta

Professor of Mathematics, Birla Institute of Technology, Ranchi, India

INDIAN MATHEMATICAL SCIENCES ABROAD DURING PRE-MODERN TIMES

    The spread of Buddhism in other Asian countries paved the way for cultural and scientific intercourse with India. The Chinese version of <u>Matangavadana</u> (ca. 200 A.D.) had shadow lengths for Hindu gnomon of 12 units, and <u>Ta Pao Chi Ching</u> (541 A.D.) had an Indian system of numeration. The <u>Sui Shu</u> (7th cent.) mentions Chinese versions of several Indian works on mathematics and astronomy. I-Hsing (ca 725) constructed a tangent-table based on Indian table of sines. Indian decimal place-value system of numerals appeared in many inscription of S.E. Asia.

    Al-Fazari who was involved in translating <u>Sindhind</u> from Sanskrit during the reign of Caliph al-Mansur (755-775), displays knowledge of three Indian values of the <u>sinus totus</u>, and al-<u>K</u>hwarizmi (ca.825) quotes two Indian values of π. The latter's work on Indian arithmetic, extant as <u>Liber Algorismi de Numero</u> Indorum, played a great role in spreading Indian numerals in Europe. Following and improving on him were several works on Indian computational methods written by Arabic, Persian, Latin, and other scholars. Later on, translations of Graeco-Arabic and Indo-Arabic works played a part in European Renaissance mathematics.

Erkka J. Maula

SF-14700 Hauho, Finland

TOPICALITY VERSUS UNIVERSALITY IN NEW IDEAS : EXAMPLES

"Great ideas" in mathematical science tend to become universal almost by definition, traversing demarcation lines of departments, cultures and periods, and eclipsing their origins. Initially, however, they bear the hallmark of thei creators which may be the only vestige of their history. It is perhaps a favoured pattern of integers, like Pythagorean triples (cf. L.B.v.d.Waerden, Geometry and Algebra in Ancient Civilizations, 1983), or even a non-discursive but successful procedure. At any event, it indicates a dialectical tension between theory-formation and concept-formation. I call it topicality, which is rather a tentati notion than an analytical tool. Even though not contributory to the growth and transition of science in any simpl manner, it nevertheless serves the historian far better than does the commonly tolerated confusion of logical an historical order. I exemplify the notion by references to a series of case-studies.

In my reconstruction of the oldest known scientific instrument ("The Calendar-Stones from Moenjo-Daro", Interim Reports I:159-170, 1984 Aachen; RWTH Aachen&IsMEO Roma), th hallmark is the pivotal ratio (2:1) which permeates the entire Indus Culture. In the reconstructed Harmony of the Spheres ("The Conquest of Time", Diotima 11:130-147, 1983 Athens), a predilection for two integers (2,3) and their powers marks out both the oldest authentic musical scale and a system of constant orbital velocities with alternative planetary periods then current that paved the way f all theories of concentric spheres (cf. E.J.Aiton,"Celestial Spheres and Circles", History of Science XIX:75-114) In Eudoxus of Cnidus' main instrument ("From Time to Pla /The Paradigm Case", Organon 15:93-120, 1979 Warsaw), later confirmed by archaeological finds, all other functions ultimately rest on its use as the first-ever measure of ang les in terms of $\tan(a/2) = q:p \Rightarrow \tan(a) = 2pq:(p^2-q^2)$, yet another example of Pythagorean triples in Greek sciences (cf. A.Szabó&E.Maula, Enklima, 1982). Finally, in the ancient analysis and synthesis ("An End of Invention", Annals of Science 38 No. 1:109-122), commitments to operations the reverse of which alone are deductive, bear the hallmark of the genius who perfected this heuristic tool.

If a moral must be extracted, it amounts to the maxim that any historian interested in the genesis of "great ideas" should keep an eye on topicality rather than universality, even though it runs counter to current principles of the pseudo-science called "rationalist history"

LAM Lay-Yong

Associate Professor, National University of Singapore, Singapore.

SOME GENERAL CHARACTERISTICS OF MATHEMATICS IN TRADITIONAL CHINA

   The use of the counting rod mechanism by the ancient Chinese led to the notational and symbolic concepts in mathematics. Computation by counting rods showed a decimal place value numerical notation in which, either a large number or a number with a decimal fraction was expressed without difficulty. In the third century, Liu Hui expressed numbers to five decimal places. The first decimal place was called fen and the numbers in this place were designated to a specified column on the counting board so that the need of a decimal point did not arise. Negative quantities were represented by black coloured rods and a blank on the counting board denoted zero.
   Generally, the methods in the texts were written in a terse fashion without detailed explanations, as they were merely instructions to be carried out for computing by the counting rod system. From commentaries and notes, the initial derivations of some of these methods were traced to a geometrical origin, where the basic concept was the manipulation of areas of varying shapes and sizes. While it was not possible to abstract a common geometrical method from the different solutions of particular problems, when these geometrical methods were transcribed for computation by counting rods, through the process of time which sometimes took centuries, patterns and similarities were perceived on the board and these led to generalisations and abstractions. In the evolution, the algebraic concept was established, where solutions were intended for sets of the same types of problems. The algorithm method emerged as a method conducive to the counting rod system. Furthermore, positions on the counting board no longer just held rod numerals but also symbolised mathematical concepts. For instance, a mathematician in the 13th century, using the tian yuan notations on the counting board, could formulate a polynomial equation of any degree from the data of a problem as easily as anyone today, using our present algebraic notations. Methods recorded for computation by the counting rod system could be adapted with certain modifications for computation by other means, such as by pen and paper. So when the methods and concepts in Chinese mathematics were transmitted to countries which did not use the counting rod system, a necessity arose in these countries for the conceptual notations and symbols to be written in a solid form. Thus, a crucial first step in providing the essential tools to build the foundation of modern mathematics was taken.

Christoph J. Scriba

Institut für Geschichte der Naturwissenschaften
Universität Hamburg

Transmission and Tradition:
A look at wandering mathematical problems

It is well known that a large number of elementary mathematical problems can be found in almost identical form in various civilizations. Ancient or medieval texts from Mesopotamia and Egypt, from China, India, Greece, Byzantium, Islamic countries and Western Europe contain numerous examples which suggest that a large proportion of such problems has been transmitted over the centuries and migrated from one cultural area or civilization to another. Other types of problems, though showing great similarities, may have been formulated independently, especially if they rose from everyday life. D. E. Smith and Vera Sanford about 1920, and more recently Kurt Vogel have begun to investigate the interrelations between such problems stemming from vastly different sources. A preliminary survey can be found in ch. 4 of Johannes Tropfke: Geschichte der Elementarmathematik, 4th ed., vol. 1: Arithmetik und Algebra, vollständig neu bearbeitet von Kurt Vogel, Karin Reich, Helmut Gericke. Berlin, New York 1980. The authors rightly emphasize (p. 513): "eine umfassende Geschichte der Rechenaufgaben.... wurde bis jetzt noch nicht geschrieben." - In my paper I shall outline what such a comprehensive history of mathematical problems might accomplish. Systematic research for the influence of transmission on tradition could, in particular, contribute to better understanding of the process of assimilation that takes place when a body of mathematical or scientific thought is crossing cultural borderlines. Due to the intrinsic difficulties of such an investigation (not the least of which are the numerous and vastly different languages in which the source material has been written) this kind of research could only be undertaken by a team of specialized scholars. For optimal results a collaboration with historians of science, of technology and of other related disciplines is indispensible.

Tatsuhiko Kobayashi, Kaoru Tanaka

Jutoku Senior High School, Kiryu, Gunma, Japan

COMPARISONS FOR THE STUDY OF THE CYCLOID BETWEEN THE WEST AND THE EAST

On the history of mathematics of modern ages in the West, it is not too much to say that the first scholar about the cycloid was G.Galilei. He discussed it with the motion of the pendulum. After his study, mathematicians or scientists found out the matters concerning cycloid in less than no time. The object of their study was not only mathematical interest but also was connected with the Astronomy and Physics. Especialy C.Hugens demonstrated that the cycloid has an isocronomism and that the evolute of a cycloid is a cycloid in his book called "Horologium Oscillotorium" (1673).

On the other hand, the first study of the cycloid in Japan appeared in "Rekisho-Shinsho" (1800) written by T.Shizuki. He wrote its book from J.Keill's "Introductio ad veram Physicum et veram Astronomiam" (1725) translated into Dutch. Later 30 years, when the Japanese mathematicians of the Edo period took up the cycloid problems all at once as a part of mathematical study, the region of its study extended from the plain to the sphere as fig. Therefore generally, it is said that they had not the concept of the motion in their study. But we can't belive old opinions, because we can point out the new facts from their notes.

"Sanpō Eari-Kan"
(1834)

Eberhard Knobloch

Prof. Dr., Technical University of Berlin (West)

EULER AND THE HISTORY OF A PROBLEM OF PROBABILITY THEORY

There seems to be an insignificance of the heritage bequeathed by Euler in the theory of probability. But when I studied in 1984 Euler's unpublished mathematical manuscripts in Leningrad, I found about 86 manuscript pages concerning probability theory and related topics. Euler especially studied a problem of probability theory which is closely connected with combinatorial analysis and which has been discussed by many mathematicians before and after him. The problem appeared originally in connection with the mathematical analysis of the game of cards "Treize" or "Rencontre". It can be expressed as follows: How many times can we arrange n numbered things in such a way that no one is in its appropriate place? The solutions of Montmort, Nicholas I Bernoulli, John I Bernoulli, de Moivre, Euler, Lambert, Waring, Laplace, are discussed which were partly unknown in the nineteenth century.

Wenlin Li, Institute of Mathematics, Academia Sinica, Beijing, China

On the Alternation of the Algorithmic and Deductive Trends in the History of Mathematics

Emphasizing the distinction between the algorithmic and deductive tradition in the history of mathematics rather than the opposition of the geometrical and algebraic approaches, this article suggests a spiral development of mathematics-the progress of mathematical science is considered to be characterised by an alternation between the algorithmic and deductive approaches as dominant trends.

The role that the algorithmic tradition played in the history of mathematics are examined in two respects:

(a). Discovering great truths. In particular, an investigation on the algorithmic idea in Descartes' "GEOMETRY" is made here;

(b). Underlying the deductive intelligence. The transition, for example, from the infinitesimal algorithms of Kepler, Fermat, ..., Newton, Leibniz to the rigorous modern analysis is viewed in this light.

Being two diffrent forms of mathematical intelligence, the algorithmic and deductive trends appear to be affected by geographic culture and milieu. The author presents a tentetive illustration of the flourish of the Indian and Chinese algorithms in ancient and medieval times. There is also a comparative analysis about Cambridge and Göttingen in the 19th. century to account for the different mathematical styles there in this regard.

Mrs DAHAN - DALMEDICO Amy

Research worker CNRS Paris

THE MATHEMATIZATION OF ELASTICITY BY CAUCHY
A problem of transmission or a problem of modelization ?

In my introduction, I point out that during the first half of the 19 th century, the développment of the mathematical physics happened around two competitions :
1) a first competition between Lagrange's and Laplace's approaches.
2) a second competition between the new mathematical analysis and the methods directly inherited from the 18 th century, like for instance, the calculus of variations.

The statement is mainly directed at the theory on Elasticity, a subject which steps beyond the divisions within physics as known in 1800, the divisions between general physics (mechanics, astronomy and optics) and specific physics. In fact, Elasticity involves both the mathematical technics of mechanics and of the internal structure of the substance.

I very briefly recall the work conducted by Sophie Germain, Poisson and Navier with respect to the two above-mentioned divisions. However, the main topic of my statement is the description of the two Cauchy's theories on Elasticity, and his lasting rallying to the mechanico-molecular paradigm.

I analyse the deeply new aspects of the first method and the connection between this first method and Fourier's style in mathematical physics. I then mention the yet not welldefined move to the second method in relation to the general situation of "Laplacian programm" at this period.

I finally emphasize that in both cases, Cauchy thought of having discovered a universal model able to represent various physics phenomenons.

Gert Schubring

Universität Bielefeld

QUELLES MATHEMATIQUES A TRANSMETTRE? L'EXEMPLE PRUSSIEN AU XIX SIECLE

Quand on discute aujourd'hui de la transformation du savoir scientifique dans les pays du tiers monde, on identifie parfois les mathématiques européennes aux seules mathématiques pures, et on accuse ces dernières de n'être pas adaptée aux besoins des pays en voie de développement. Or, on peut clarifier le statut des mathématiques pures et leur apport au développement d'un pays en comparant les mathématiques en France et en Prusse au 19è siècle. Au début du siècle ces deux pays étaient sous-développés si on compare leur structure industrielle à celle de l'Angleterre à la même période, cependant que les sciences exactes et les mathématiques étaient plus avancées en France. A la fin du siècle, les rapports avaient profondément changé: c'était en Prusse que les sciences étaient maintenant à l'avant-garde, et l'économie de ce pays était par ailleurs tout à fait compétitive avec l'économie anglaise.

Si on analyse la situation, on trouve que la conception des rapports entre sciences pures et sciences appliquées était tout à fait différente en France et en Prusse, ainsi que leur institutionnalisation: en France, le fameux modèle de l'Ecole Polytechnique se modifie en 1810, à la suite de fortes pressions sociales pour développer un savoir appliqué; elle se transforme en une institution de formation professionnelle, dans un sens étroit, elle ne devient pas le lieu d'une recherche institutionnalisé, destiné à transmettre le savoir à des larges couches sociales. En Prusse, les protagonistes les plus ardents de la science pure: C.G.J.Jacobi et A.L. Crelle envisagent - en accord avec le ministère de l'instruction - d'établir à Berlin un Institut Polytechnique, mais pour y former des enseignants, non des ingénieurs. La conviction générale, aussi bien des chercheurs que des transmetteurs de savoir (par exemple les Grassmann), est en effet que la transmission la plus large des mathématiques pures est une précondition pour encourager les applications.

De fait, les sciences exactes sont alors fortement promues en Prusse, et non pas méprisées, comme on le déduit souvent des attaques de Liebig.

Néanmoins, il n'existe pas pour autant a priori un équilibre stable entre les fondements et les applications d'une science. Le dernier tiers du 19è siècle montre, en Prusse, le danger que se crée une coupure dans ses rapports mutuels et que naisse un système de sciences pures isolé et fermé sur lui-même.

Ainsi, il ne s'agit pas seulement de discuter du rôle que joue pour le développement le savoir à transmettre, mais aussi les mentalités à travers lesquelles le savoir transmis est transformé, dans le cadre de structures institutionnelles données.

Scholz, Erhard

Instructor, Bergische Universität Gesamthochschule Wuppertal, FRG

THE USE OF PROJECTIVE GEOMETRY IN 19-th CENTURY STRUCTURAL ENGINEERING METHODS

When construction of edifices became standardized about the middle of the 19-th century, graphical statics arose as a system of scientifi methods to analyze stability of structures. It was particularly well adapted to the analysis of determined frames used as structural components of bridges and halls. Different attempts were made to introduce theoretical concepts from advanced parts of contemporary mathematics into graphical statics. C.Culmann and, independently, W.M.Rankine and J.C.Maxwell developed different duality ideas in this conte

Culmann detected a duality relation of a certain kind between force and funicular polygons of graphical statics and made it the core of a broad attempt to theorize engineering statics on the background of methods and concepts from projective geometry. The development of hi view between 1866 and 1875 by and by pushed back those graphical ope rations which implicitly carried a vectorial structure, and led to a substitution by projective concepts. This shows that his style of mathematization was much more an attempt to transfer mathematical concepts to statics according to a rather "early" postulated program of theorization, than it was a move to make explicit those relations which were already present implicitly in the graphical operations.

Rankine's and Maxwell's approach, on the contrary, was more receptive to methods from different mathematical disciplines (abstract relation concept of duality, more cautious link to projective geometry, experimentation with concepts from the topological theory of surfaces) a thus was not bound to a strict program of theorization of a more or less a priori character. It led to a concept of duality which linked up with the modes of construction used by the practitioners of graph: cal statics and, at least partially, corresponded to the practical criteria of acceptability for theoretical innovations uphold by the engineering community.

The differences in the use of duality concepts was characteristic fc different styles of theorization, which, of course, engineers responded differently to. Therefore the efforts to transmit concepts and methods from projective geometry to graphical statics seems to be a rewarding object for a case study of conditions, forms of development and problems in the transmission of concepts from theoretical mathematics to an applied discipline.

Ref.: E.Scholz 1984. Projektive und vektorielle Methoden in Culmanns Graphischer Statik. NTM - Schriftenreihe Geschichte Naturwiss., Technik u. Medizin 21, 49-64.

Karine Chemla

CNRS

COMPARAISON DE PROCEDURES ET APPLICATION A DES PROBLEMES DE TRANSMISSION

Les solutions que les textes mathématiques anciens et médiévaux proposent auz problèmes qu'ils renferment sont souvent énoncées sous forme d'algorithme, de procédure de calcul. Notre travail de recherche vise à éprouver l'hypothèse selon laquelle ces procédures ont pu être l'objet, le support d'un travail mathématique.

Nous avons été amenées pour cela à considérer, pour un ensemble donné d'opérations (division, extraction de racine...), les structures, dans divers traditions, des ensembles de procédures qui les réalisent; en nous appuyant sur la nature des ressources mises en oeuvre par ces procédures, sur les modalités de cette mise en oeuvre, sur les relations établies entre leurs textes, nous pouvons définir, pour chaque tradition, des classes homogènes de procédures, lesquelles renvoient à une structure de l'espace des opérations. Nous nous proposons de montrer comment d'une tradition à conclusion de développement autonome. Dans le cas contraire, nous illustrerons le fait que l'analyse des variantes d'une procédure, des transformations qu'elle subit par transport fournit des matériaux pour l'analyse de sa structure dans chacun de ses états.

Tamotsu MURATA

Professor, Rikkyo University, Tokyo, Japan

TRANSMISSION OF CANTOR'S ABSTRACT SET THEORY IN FRANCE UP TO 1905

Our aim is to give a picture of how French mathematicians responded to each step of advancement of the Cantor's abstract set theory and, through it, to make a remark on a general feature of transmission of a new theory into other mathematical traditions.
1) A remark on the formation of Cantor's theory
We distinguish two trends in its formation. One of them started from analysis as an <u>index</u> n of a derived set of points (1872) and, through <u>symbols</u>, ω etc., of extended derived sets (1880), arrived at their transformation into <u>transfinite ordinal numbers</u> (1883). The other started from number theory as a <u>distinction</u> between the countable set and the continuum (1874) and, through quantification by p<u>ower</u> of infinite sets (1878), arrived at a <u>numeration</u> of powers (<u>cardinal numbers</u>) by means of ordinals. It is here that these two aspects were brought together into the abstract set theory.
2) Transmission in France
While Cantor's theory was coldly neglected in Germany, it was noticed in France and Italy. Especially on 1883, when Cantor's masterpiece [M-4<s>5</s>] was published and some of his early papers were translated into French, J.Tannery wrote a very long review in <u>Bull. Soc. Math. Fr</u>. However, it is remarkable that he considered the theory as a theory of real numbers, neglecting the most original parts of it; for example, Continuum Problem. In retrospect, one might say that J.T. could not truly understand it, but it is my opinion that one should admit that this attitude was natural and inevitable, considering the French tradition of mathematics. French mathematicians' main concern at that time was, it seems, in analysis. Their conception of function, e.g., was a concrete one; there was little room for an abstract concept of one-to-one correspondence.
This would also explain the attitudes of E.Borel, R.Baire and H. Lebesgue, diciples of J.T., on the occasion of Zermelo's abstract proof of well-ordered theorem [<u>Cinq lettres</u> ..., 1904]. Among them R.B. advocated the most fundamental criticism against the very idea of set theory. On 1904 he wrote that one can never find out any common measure (such as <u>power</u>) between the totality of natural numbers and the continuum.
Incidentally, "effective" mathematics by Borel and others declined once after 1930's, but is rising again in the recent studies of the foundations of mathematics. This may also be an example of wax and wane of a mathematical idea in history.
3) An additional remark
An established mathematical truth is universal, but a mathematics in a developing stage may not necessarily be so regarded. It is, after all, a product of human endeavor, and thus can not be free from particular traditions of human culture. And a transmission of heterogeneous elements would possibly be a d riving force for math.

Inés Harding, Emeritus Professor, UTE, Santiago, Chile. Visiting Professor, IDM, USP Lateinamerika, Univ. Bielefeld, F. Germany.

TRANSMISSION OF MATHEMATICS IN CHILE

The author, in the recent years, is working in order to analize the development of Mathematics in Chile from XVI C. up to present. The local History of a Science has begun iniciated to develop only in the last few years, according to T. Kuhn. In the work is presented an historical review. We have in a suscint form, the Institutionalization of Mathematics in its teaching and research. It shows us that the History of Mathematics in Chile is, almost completely its History of Mathematical Education. It presents the generation and evolution of chilean mathematical thought. It allows us to know the different stages, starting from the indigenous intellectual concepts, which during the pre-columbian period were of an entirely primitive character.
It was only eighty years before the arrival of Spanish, that the Incas came into contact with the Araucan Indians. They brought with them the decimal system of counting with the concepts of hundred and thousand. They also transmitted the use of the "quipu"__an instrument for counting.
The medioeval scholastic is transfered at the beginning of the spanic era, while in the western civilization the renaissance scientific revolution and the enlightenment are take place. Specifically in Mathematics, the aristotelian and medioeval euclidean thoughts are transfered. This is the case in all Latin America. In Chile started with "Liberal Art Courses, in 1591. Courses at the Pontifical Universities founded in 1619 and 1621. Continuing with mathematical Courses at the "Real Universidad de San Felipe" iniciated in 1738. Teaching was done in Latin with spanish books. In 1798, with the foundation of the "Real Academia de San Luis" were taught the first Courses of Mathematic in spanish. In 1813, with the Independence of Chile, the "Instituto Nacional" was founded and the systematic teaching of Mathematic in spanish was begun. In 1833, the first mathematics books were published in Chile. Gorbea, a Spanish who had lived many years in London and Paris, made many translation of french text: Francoeur, Leroy, Allazy, etc.. He carried through the reforms in mathematics proposed by the french engineer Lozier.
In 1889, the "INSTITUTO PEDAGÓGICO" was founded to prepare high school teachers. German specialized Scientists were hired. They wrote the mathematics textbooks for the lyceum and the University. The formation of researchers in Mathematic began in 1958, although in a unsystematic way. Only in 1962 did it become formalized with the creation of the degrees of Licenciado and Magister in Mathematic, latest the doctorate. For these studies were hired northamerican, french and german Mathematicians.

Ubiratan D'Ambrosio

Professor of Mathematics, Universidade Estadual de Campinas, Brazil.

"Marginal" mathematical practices and new possible ways of teaching Mathematics.

Our subject lies on the borderline between History of Mathematics and Cultural Anthropology. We may conceptualize ethnoscience as the study of scientific and, by extension, technological phenomena in direct relation to their social, economic and cultural background. There has been much research on ethnoastronomy, ethnobotany, ethnochemistry and so on. Not much has been done in ethnomathematics which remain as "marginal" in the body of mathematical knowledge.

Much has been said about the universality of Mathematics. This seems to become harder to sustain as recent research, mainly carried on by anthropologists, show evidence of practices which are typically mathematical such as counting, ordering, sorting, measuring and weighing, which are carried on a radically different way than those which are commonly taught in the school systems. This encouraged further studies on the evolution of the concepts of mathematics in a cultural and anthropological framework. We consider this to have been done only to a very limited and we might say timid extent. On the other hand, there is a reasonable amount of literature on this by anthropologists. The bridge between anthropologists and historians of culture of mathematicians is an important step towards recognizing different modes of toughts which lead to different forms of mathematics, which we may call ethnomathematics.

These remarks invite us to look into the History of Mathematics in a broader context, so to incorporate in it other possible forms of Mathematics. But we go further on these considerations in saying that this is more than an academic exercise, since its implications for the Pedagogy of Mathematics are clear.

We would like to insist on both the broad conceptualization of what is Mathematics and which allows us to identify several practices which are essentially Mathematics in their nature. And also we presupose a broad concept of _ethno_, which includes cultural identified groups in their jargons, codes, symbols, myths and even specific ways of reasoning and inferring. Of course, this comes into a concept of culture which is the result of an hierarquization of behavior, from individual behavior through social behavior and leading to cultural behavior. This relies on a model of individual behavior based on the cycle ... reality → individual → action → reality. The ceaseless cycle is the basis for the theoretical framework upon which we base our ethnomathematics concepts.

Lee Peng Yee

Department of Mathematics, National University of Singapore

The role of SEAMS in the transmission of mathematics in SEA

We give an account of the recent history of how the Southeast Asian Mathematical Society (SEAMS) played an important role in the transmission of mathematical knowledge in Southeast Asia. Amongst others, the society was instrumental in establishing a graduate programme in the Philippines, and in stimulating research in graph theory in the region. Hence the transmission of mathematical knowledge among the countries in Southeast Asia was made possible. It is history in the making, and I believe it is worthwhile to put on record what has happened in the immediate past and what will have happened in the near future.

GALLUZZI, PAOLO. Professor, History of Science, University of Siena,
                  Italy
Director, Istituto e Museo di Storia della Scienza, Florence, Italy

The history of science in Italy: problems of documentation

　　　In recent years, several projects concerning the problems of
documentation in the history of science have been undertaken in
Italy. The focus has been the access to primary sources.
　　　Four major projects have been promoted by the Istituto e Museo
di Storia della Scienza in Florence. The <u>Bibliografia Italiana di
Storia della Scienza</u>, is to be published annually. The first two
issues of the <u>Bibliografia</u>, already published, list contributions
to the history of science printed in Italy during 1982 and 1983 (for
a total of approximately 2500 titles). The <u>Bibliografia</u> has ample
and detailed indices which make consultation extremely easy. The
future issues of the <u>Bibliografia</u> will contain abstracts and a list
of dissertations in the History of Science submitted to Italian
Universities. The second undertaking promoted by the Florentine
Instituto bears the title <u>Archivio della corrispondenza degli
scienziati italiani</u>. The goal of this research is the systematic
gathering of microfilms of the scientific correspondence relevant
to the Italian history of science. The material will be catalogued,
indexed and, eventually, the most significant groups of letters will
be be published. The whole program is run by computer. The first
volumes of the <u>Archivio</u> have already been printed. The third
undertaking is devoted to a survey of the scientific instruments
preserved in Italy. A special and very complex program has been
devised in order to process the data. As to the fourth project, it
consists in the preparation of a <u>Guida alla Storia della Scienza</u> in
<u>Italia</u>. This directory will list Italian scholars interested in the
history of science, the major Institutions and Periodical publications in this field and, finally, the academic positions in the
history of science available in Italian Universities.
　　　Attention will also be paid to research on problems of documentation promoted in this field by other Italian institutions or
groups. The recent foundation of the Italian Society of the History
of Science may contribute in the future to open new perspectives
in this direction.

Andreas Kleinert

Universität Hamburg, Hamburg, Federal Republic of Germany

ARCHIVALIEN ZUR WISSENSCHAFTS- UND TECHNIKGESCHICHTE IN DER BUNDESREPUBLIK DEUTSCHLAND

Es gibt in der Bundesrepublik Deutschland keine zentrale Stelle, die sich um die Erhaltung, Aufbewahrung und Erschließung von Quellen zur Geschichte der Naturwissenschaften und der Technik bemüht. Gelehrtennachlässe, Briefwechsel und andere handschriftliche Quellen werden in zahlreichen Archiven, Bibliotheken und Museen aufbewahrt, wobei die Wahl des Aufbewahrungsortes oft auf Zufälle zurückzuführen ist. Wichtige Nachlässe von Naturwissenschaftlern und Technikern befinden sich in den Sondersammlungen des Deutschen Museums (München), in der Staatsbibliothek Preußischer Kulturbesitz (Berlin) und in den Handschriftenabteilungen einiger Universitätsbibliotheken.

Die Aufbewahrungsorte einzelner Gelehrtennachlässe werden durch die Verzeichnisse von Mommsen (Die Nachlässe in den deutschen Archiven) und Denecke (Die Nachlässe in den Bibliotheken der Bundesrepublik Deutschland) nachgewiesen, die kurz vorgestellt werden. Einzelne Briefe können über die weithin unbekannte Zentralkartei der Autographen in der Handschriftenabteilung der Staatsbibliothek Preußischer Kulturbesitz ermittelt werden.

Es wird erläutert, welche Möglichkeiten bestehen, um Gelehrtennachlässe oder einzelne Autographen aufzufinden, die sich in Privatbesitz befinden, und es werden die urheberrechtlichen Bestimmungen erläutert, die beim Veröffentlichen ungedruckter Quellen wie Briefe, Manuskripte usw. zu beachten sind.

Boris V. Levshin

USSR Academy of Sciences. Archive. Moscow. USSR.

The problems of using documentation for studying on science history in the institutions of the USSR Academy of Sciences

The most valuable sources for studying science history is documentation which is the result of the activity of the USSR Academy of Sciences and its bodies. According to its origin it consists of two parts: documentaly funds of institutions and funds of scientists, which interact each other and help to found informational connections of documents. Optimum using the archival funds of the USSR Academy of Sciences for studying on science history is caused by a number of factors:

The first one is connected with the universal purposes of scientific investigation, resulting in the huge bulk of documentations for the solution of problems.

The second factor, determing the complexity of investigation is the specification of the scientific documentation including repetition in documentary materials.

The third one works out valuable and scientifically wellfounded methods of estimating documental information paying attention to the process of its movement, transformation, succession in the document systems, level and place as well.

The pecularity of using the documental funds of the USSR Academy of Sciences for scientific investigation on the modern step of development is:

1). Scope and scale of the scientific activity during technological revolution, the high rate of growth of investigations.

2). Appearing new nontraditional kind of specific documentation for fixing up scientific experiments and involving it in the investigation process.

3). Applying the computing system of information processing providing succession of information value of documents, created on the paper and machine information carrier constituting interacting elements of the documental system.

S. A. I. Tirmizi

SCIENTIFIC ASSOCIATIONS OF THE RAJ

The purpose of this paper is five-fold. Firstly it purports to identify the causes of non-institutional traditions of Indian science in the pre-colonial period when caste solidarity, racial zeal, and commercial interests encouraged propensity towards unities which were usually non-secular in nature. Secondly it aims at identifying the new pressures and competitive circumstances prevailing under the Raj, which compelled the old unities to adopt the new techniques of associations under the inspiration of similar institutions in Britain. Thirdly it attempts to trace out the motivation in the establishment of these associations by the persons who were united by common education, common aspirations, common skills, and common disciplines. The Indians who converged on these new associations were men who had learnt the scientific idioms of the Raj. It is, therefore, not surprising if these new associations were first organised in the presidency capitals of Calcutta, Bombay, and Madras. Fourthly it intends to study the growth of these specialised and general associations in the context of politico-economic requirements of the Raj. Fifthly it purports to analyse the aims, objects, and activities of these associations with a view to determining the vital role they have played in the propagation and dissemination of science and also in fostering scientific consciousness through the length and breadth of a vast country of subcontinental dimensions.

Rudolf H e i n r i c h

Keeper of Special Collections, Deutsches Museum, Munich,
Germany

Pictorial sources - an underestimated means of historical research

A general survey of the different kinds of pictorial sources useful to historians - e.g. drawings, etchings, photographs, documentary films - is followed by a discussion of access-modes, ranging from card catalogs and printed documentations to halftone microfiches and videodiscs.

In order to demonstrate how pictures can be superior to written texts especially in the history of technology, and, on the other hand, how much experience in the technical, social and auxiliary sciences is needed to enable satisfactory interpretations when captions are missing and/or uncommon drawing-techniques are employed, several examples from the archives of the Deutsche Museum are presented together with interpretations of the late professor Friedrich Klemm, one of the leading historians in this field.

Multhauf, Robert P.,

Smithsonian Institution, Washington, DC, U.S.A.

DOCUFACTS: OR WHAT IS THE USE OF THE COLLECTIONS OF THE MUSEUM OF SCIENCE AND TECHNOLOGY.

Although there are a few relevant university museums of some antiquity (e.g. Oxford, Utrecht), the museum of science and technology essentially originated in the international exhibitions of the 19th century. The inspiration, like that of the exhibitions themselves, was largely an enthusiasm for technology and a belief in the inspirational value of its material remains, especially where they could be associated with famous inventors. These motivations have passed, and the "science museum" of today is commonly a mixture of school experiments, often animated, with a show case for commerical or military hardware. It collects nothing.
The older collections of the material relics of science and technology are still with us (with a few of this type where new nations have come into being. It is the purpose of this paper to consider the usefullness of their holdings.
Some museum objects, notably tools, have been fundamental in studies of anthropology and archeology. There are cases (e.g. microscopy) where the history of science is dependent upon the survival of examples of early apparatus. Other examples reveal that "obsolete hardware" has obviated the need for its reinvention. Museum objects have on occasion proven important in litigation over priority of invention.
These and other examples will be discussed.

Spencer R. Weart

Center for History of Physics, American Institute of Physics, New York

ORAL HISTORY OF SCIENCE -- METHODS AND MISTAKES

Oral history has long been accepted by general historians, yet historians of science still have much to learn about this method. The pioneers have been historians interested in physics and astronomy, who have accumulated thousands of hours of oral history interviews in large collective efforts and in numerous individual projects. They have found that oral history of science poses special difficulties and opportunities.

Scientists may insist that everything of value in their lives can be found in the rationalized record of their published scientific papers. In their memories too they are likely to rework their scientific career to put it in a logical "textbook" order. Also, concern for priority in discovery may dominate the interview; a permanent place in history is a main personal goal of many scientists. Interviews will be most reliable when they seek information that does not trespass on concerns of logic and priority.

Some historians believe that paper documentation such as correspondence is more reliable than interviews. However, each type of data has its own pitfalls and uses. Even unreliable interviews can give clues to areas worth researching and to ways of interpreting written data. Furthermore, interviews can give vital information that was never put on paper; by cross-checking interviews against one another, many historians have verified such facts. This procedure becomes essential for studying the post-World War II era, when written communications become more voluminous and less frank, and when the increased complexity and size of the community call for guidance from the people who actually lived in it.

Oral history will lead one astray unless modern techniques are used. One must prepare extensively in advance, typically a few days of research for each interview; one must support one's own memory with a tape recorder. Transcription into written form may be avoided for a short interview focussed on a narrow topic (written notes and an abstract will suffice), but transcription is necessary for biographical interviews meant to be used by future historians. In either case the historian should record whether the person interviewed wishes to restrict use of the interview.

Otis Dudley Duncan, University of California, Santa Barbara, CA, U.S.A.

NOTES ON SOCIAL MEASUREMENT

In my <u>Notes on Social Measurement</u> (New York, 1984), I suggested that the social history of measurement might be extended to include social measurement, taking advantage of Hunter Dupree's thesis that "measures are a key to the needs of past and present societies." I have tentatively identified the main inventions that seem to underlie modern practices of social measurement and have attempted to trace their historical origins and elaborations in a superficial way. They include: (1) voting; (2) counting people, as in a census; (3) valuing goods and services in units of a standard commodity; (4) defining and labelling social ranks and degrees; (5) appraising the quality of persons and performances with some kind of contest or examination; (6) designing appropriate rewards or punishments for achievements or transgressions. These six inventions were made in antiquity, some in prehistoric times, and for several of them we can find evidence that the ancients struggled with problems that face the same kinds of measurements that are being made today, for example (1) public opinion polls, (2) national censuses, (3) systems of economic statistics, (4) studies of social stratification, (5) educational testing and the keeping of athletic records, and (6) quantification of the seriousness of criminal offenses. I identify, as well, several inventions made in the "scientific era," often by persons identified, broadly, as social scientists: psychophysical scaling, index numbers, summaries of statistical distributions, methods of measuring utility, and quantifications of properties of social networks. One key ingredient of modern social measurement, the concept of probability and application of the calculus of probabilities, straddles ancient and modern times. Although the ancients had no concept of probability as such, they appreciated the aleatory element in social life and designed sophisticated procedures for random sampling and random assignment that have not been as carefully studied by students of the history of probability as they should be.

Although most of my remarks concern the opportunities for historical scholarship in tracing the genesis and development of measurement in social affairs, there is also the important inverse problem of assessing the impact of measurement practice on the social process. Patricia Cohen's monograph on the spread of numeracy in early America advances our understanding of the way in which the demand for measurement grows and how numerical information and reasoning came to supersede other kinds of political argument.

To understand why and how societies carry out their repertoires of social measurements is to know a great deal about social structure and the processes of social change.

Heinz Ziegler

Universität GH Siegen, Siegen, Germany(BRD)

THE WELL-PROPORTIONED MAN AS RELATED TO PROTAGORAS' HOMO MENSURA THEOREM

We refer to man as a measure and, more specifically, to his limbs as measurable quantities, since the human limbs have of old been used to define measures and they have been understood to be interrelated in specific ratios.

In occidental cultures, linear measures were related in specific proportions to the height of the body which at the same time was the largest measure derived from the human body - the so-called fathom. This largest dimensional unit, which the Greek called Orgyia, was subdivided into 4 ells of 24 fingers each, or 6 feet of 16 fingers each, very often also into 8 spans of 12 fingers each. The palm or handbreadth comprises 4 fingers the finger (more correctly the breadth of the finger) as the smallest dimensional unit being given the figure 1, thus representing 1/96th of the height of the body.

As interrelations can be understood and described only with the aid of figures, the human proportions by necessity are related to figures, and the human limbs, used to define measures, were inseperable from their associated figures. Height and breadth of a human being are likewise related in a specific proportion which can be said to follow a regular pattern. This proportion had to be the right one as the words in the Book of Proverbs (Salomon 11,21) - "Omnia mensura et numero et pondere disposuisti" - have been accepted as a general truth.

Even in pre-Christian times, figures were understood to exist a priori, i.e. figures had the precedence of matter. And if a measure was to be "rightful", it had to represent the right ratio, which means it had to contain the "right" figure.

To regard man as "the measure of all things" could imply that man is built in compliance with the "right" ratios, i.e. that the limbs have the ("right") figures assigned to them and this in the right proportion.

Ellen Z. Danforth

Curator, Streeter Collections, Yale University

THE STREETER COLLECTION OF WEIGHTS AND MEASURES: A "NEW" RESOURCE FOR METROLOGISTS

Dr. Edward Clark Streeter, who began collecting weights and measures in 1923, was fascinated by the evolution of man's scientific curiosity. He based his collecting on the belief that metrology is the foundation of all sciences. It is the common "thread" that unifies his collection of an unprecedented array of scales, weights, and measures of volume and length. The Streeter Collection of Weights and Measures at Yale University's Medical Historical Library is the culmination of one man's lifelong quest to document science and history.

Although Streeter knew very little about weights and measures when he started collecting them, he was a zealous collector driven by, as he said, "that inner compulsion to acquire and accumulate." It was this instinct, more than anything else, that enabled Streeter to gather together what is one of the most comprehensive metrological collections in the world.

The several thousand artifacts encompass a broad range of material: photographs, metrological ephemera, a library of literature from the sixteenth century to the present, and of course, the weights and measures themselves.

The weights of the ancient Classical World are the most important holdings because nearly every type is represented and they are all in good condition. Separately, the Egyptian, Babylonian, Assyrian, Greek, and Roman weights each form an almost complete series.

Other important holdings include the French "poids de ville", possibly the largest collection outside France or Belgium, and the set of nested cup weights, which at seventy-five pieces, is one of the most comprehensive museum-owned collections in the world.

With the exception of the nested weights, whose descriptions are currently being readied for publication, little in the Collection has ever been systematically studied or published.

When Dr. Streeter presented his collection to the Medical Historical Library in 1941, he was made Curator of Museum Collections. He continued to collect until his death in 1947, but unfortunately left no formal record of his acquisitions. Subsequently, several attempts were made to catalog his collection.

Now, finally, a complete card catalog exists, and, for the first time, the information is accessible. There are also new interpretive exhibits, and some improvements have been made in the storage of the collection.

The purpose of this presentation is to call attention to the availability of this study collection to metrologists.

Prof. Dr. Karin Figala,

Technische Universität München, München, W.-Germany

NEUE ENTDECKUNGEN ZUM LEBEN UND WERK DES ALCHEMISTEN MICHAEL MAIER
(1568-1622)

Die Auffindung eines Jugendbriefes und die Entdeckung dreier bisher der Forschung völlig unbekannter Baseler Drucke aus dem Jahre 1596, sowie die Bearbeitung des Widmungsbriefes der medizinischen Baseler Dissertation geben erstmals Auskunft über die familiäre Herkunft, über Einzelheiten der Studienausbildung und nicht zuletzt über die Dichterleistung Michael MAIERs.

1. Der Brief ist insofern besonders bedeutungsvoll, als es sich hierbei um das älteste heute bekannte Autograph MAIERs handelt. Das Schreiben stammt aus dem Jahre 1590 und ist an den berühmten Mäzen der Wissenschaften, den holsteinischen Adeligen Heinrich RANTZAU (1526-1598) gerichtet, der unter anderem auch den bedeutenden Astronomen Tycho BRAHE (1546-1601) unterstützte. Abgesehen davon, daß die Förderung MAIERs durch RANTZAU der Forschung völlig entgangen ist, werden in diesem Brief die bisher nur vagen Spekulationen über die Herkunft MAIERs durch konkrete Angaben des Vor- und Zunamens, des Berufes und der Stellung seines Vaters gründlich korrigiert. Die hier vom jungen MAIER gezeigten Proben humanistischer Kenntnisse sind außerdem ein wichtiges Zeugnis für den späten Renaissancehumanismus des dänischen Holstein, der bekanntlich von RANTZAU in den "cimbrischen Norden" gebracht worden war.

2. Einem weiteren, bislang ebenfalls völlig unbekannten Studienförderer aus Holstein, dem Hofarzt Matthias CARNARIUS, ist MAIERs medizinische Dissertation von 1596 gewidmet. Dieser hochverehrte Freund hatte MAIER offensichtlich das Studium in Basel empfohlen, da die dortige Universität zur damaligen Zeit dank mehrerer Professoren einen hervorragenden Ruf insbesondere in den Naturwissenschaften und in der Medizin genoß.

3. In die Zeit unmittelbar nach MAIERs Doktorpromotion vom 4.Nov. 1596 in Basel gehören die drei der Forschung bis jetzt ebenfalls entgangenen Druckschriften. Sie liefern erstmals Belege für das rein dichterische Werk des "Poeta Laureatus Caesareus" Michael MAIER. In lateinischen Gesängen verschiedenen Versmaßes feiert er dort einige seiner mitpromovierten Kommilitonen, und in einem öffentlichen Musenspiel unter wohl von ihm selbst komponierter Musikbegleitung läßt er die neun frischgebackenen Doctores den gemeinsamen, verehrten Doktorvater, den weltbekannten Anatomen, Botaniker und Humanisten, Caspar BAUHIN (1560-1624), als Präses und Apollo einer Mediziner-Dichtervereinigung preisen. Die Geschichte der Universität Basel kennt bisher keine ähnlichen Beispiele von akademischen Dichtersozietäten oder von öffentlichen poetischen Schauspielen mit der Lorbeerbekränzung der promovierten Studenten durch den Professor Bemerkenswert ist weiterhin, daß in diesen allegorischen Poesien von 1596 bereits denselben mythologischen Gestalten Bedeutung beigelegt wird, die in den etwa 20 Jahre später publizierten alchemischen Werken MAIERs eine wichtige Rolle spielen.

SZULC Halina

Assistant Professor - Polish Academy of Sciences
Institute of Geography and Spatial Organization

The application of historical metrology in genetic research on regular villages in Poland. The case study of Silesia and Pomerania

As a result of the application of modern analytical methods particular attention is now paid to the use of historical plans of villages as source materials for studying the genesis and transformations in rural settlement. Great progress has been achieved following the introduction of the genetic-metrological method, worked out by Swedish scholars (D. Hannerberg and his followers. The method aimes the discovery of measures used in the village, and subsequently, on the basis of the knowledge of chronology of the system of measures and the spatial model of the village at a given time, it is possible to fix the period when the village was founded and the principles of its measuring. The application of the genetic-metrological method to the study of the layouts of regular villages in Silesia and Pomerania has yielded very interesting results, since it makes it possible to establish the principles of their layouts at the time of foundation. In case when no historical data are available the method enables the researcher to differentiate a regular, measured village from an irregular one, which has evolved spontaneously.
Metrological measurements make it also possible to discover certain analogies in the layout of regular villages in Silesia and Western Pomerania (Hinterpommern and those in Brandenburg, southern Sweden and England. The author has applied the method in her own research, i.a. in: H. Szulc, Osiedla podwrocławskie na początku19w (Sum.: Suburban settlements in the vicinity of Wrocław at the beginning of the 19th century),Wrocław-Warszawa-Kraków 1963, 108 + 4 maps.
H.Szulc, Typy wsi Śląska Opolskiego na początku 19w. i ich geneza (Sum.: Types of rural settlements of Opole-Silesia at the beginning of the 19th century and their origin), Prace Geogr.IG PAN, 66,Warszawa 1968,107+4maps
H.Szulc,Studies on the Silesian Village in the Light of plans from the beginnings of the 19th c.Kwartalnik Hist. Kultury Materialnej,V.16,No.4,1968,pp.621-639.
H.Szulc,Regular green villages in Western Pomerania,Geographia Polonica,V.38,1978, pp.265-270.
H.Szulc, Morphogenetic Typology of Villages and their Regulation in Poland: A case Study from Silesia and Pomerania,Villages,Fields and Frontiers.Studies in European Rural Settlement in the Medieval and Early Modern Periods,ed.B.K.Roberts,R.E.Glasscock,BAR IS 185,1983, pp-203-215.

Denys Vaughan

Science Museum, South Kensington, London SW7 2DD, England.

LENGTH AND TIME UNITED: THE ATTEMPT TO USE THE PENDULUM AS A NATURAL STANDARD OF LENGTH

A natural standard based on some invariable property of nature has intellectual attractions in addition to the practical advantages of a standard which does not deteriorate and which can be recovered at any time. The use of a pendulum, beating seconds, as a natural standard of length was proposed in the 17th century. The improvements in pendulum measurements which occurred in the 18th century led Talleyrand to suggest that it might form the basis of the new system of measures to be introduced in France following the Revolution. The Academy of Science rejected Talleyrand's proposal and based the metre on the size of the earth but the pendulum was to be used to recover the unit should the metre bar be destroyed. The British Government adopted a similar approach in 1824 when they passed the Act which introduced the Imperial System of weights and measures. The yard was retained as the unit of length and based on a material standard but in the event of the yard becoming lost, defaced or otherwise injured it was to be restored by comparison with a seconds pendulum. In 1834 a fire in the Houses of Parliament destroyed the standards and a commission was appointed to consider the steps to be taken for their restoration. They rejected the proposal contained in the Act of 1824 and recommended that the yard should be restored by comparison with existing material standards. It seems likely that a similar course of action would have been adopted in France if the metre bar had been destroyed.

This paper will examine the reasons why the pendulum was at first accepted as a natural standard and then rejected.

Harald Witthöft

Professor, Universität GH Siegen, Siegen, Germany(BRD)

GOLD, SILVER AND SOME BASIC ELEMENTS OF THE NORTHERN
EUROPEAN WEIGHT-SYSTEM SINCE THE MIDDLE AGES -
SOURCES AND INTERPRETATION

   A great number of surviving objects testify the existence of an elaborate weight-system in Northern Europe since Franconian times. The finds and relics
- differ regionally in frequency and type,
- are more or less abundant in certain periods, and
- belong to a great variety of field of application.
But it still lacks a definition of the elements of a basic structure underlying their standards, usage and geographical and chronological distribution. Attemps in this direction have been made, but with little attention to a wider scope of surviving objects.
   The thesis to be debated is, that the older system of weights and measures has to be linked to a rational system of numerical ratios, which have their origin and meaning in natural scientific facts like density etc.
An example is to be found in weighing and calculating gold and silver in the middle ages and early modern times. The results shed some light on the kind of weights used in trading certain goods on certain routes by certain means of transport till the 19th century.
   The material basis supporting the argument exists in: Franconian pennies, small weights from viking-times Scandinavia, Cologne pennies and Rhenish guilder, the French 'pile de Charlemagne', the English tower pound, the Lüneburg Markpfund, the Prussian(Cologne) Mark, the metric pound.
   The means of interpreting the monetary weights as an inroad into the early weight-system in its widest sense is provided by the medieval way of calculating and paying "inter aurum et argentum" or "in auro et argento", i.e. in gold and silver. It needs briefly to be shown, that and how this calculation was done on the basis of firmly established ratios, the alteration of which only occured at crossroads of history - e.g. in the reign of Charlemagne (793/94).
   There is opportunity to present the thesis, that the basic reason for certain constant ratios in weighing and valueing gold and silver rests with their dependence on their specific weight(density) - 1xAu : 2xAg = 24 : (2x12 1/2) to give an exemple.
   The change of weight standards within the monetary system since ancient times and in selected countries is reflected in certain units: e.g. 327,45o g(Roman), 4o8,24o g(Charlemagne), 349,92o g(tower pound), 233,887 g (Mark Cologne), 244,7529 g(poids de marc Paris).

Dr. Ronald Edward Zupko

Professor of History, Marquette University, Milwaukee, WI U.S.A.

TYPES OF METROLOGICAL UNIT VARIATIONS IN EUROPEAN METROLOGY PRIOR
TO METRICATION

The many weights and measures systems employed throughout
western Europe during the Middle Ages were introduced and developed
by Germanic and Celtic peoples who fell heir to the western provinces following the breakdown and collapse of Roman power during
the fourth and fifth centuries. Characterized generally by confusion and complexity, and dominated largely by custom and tradition, medieval weights and measures evolved on local or regional
bases and were geared to needs, especially during the Early Middle
Ages, that were significantly different from those found in the
earlier Roman world. During most of the Republican and Imperial
eras, Rome enforced a standardized system of weights and measures
in Italy and the provinces to insure the proper functioning of its
sophisticated political, legal, military, commercial, and urban
institutions, and in so doing added yet another dimension to the
universality and unity of Roman life.

This unity, precision, and standardization -- not to be witnessed again in western Europe until the creation and dissemination
of the metric system in the modern world -- came to an end during
the turmoil and violence of the Early Middle Ages. Native metrol-
ogies, long laying dormant in the countrysides or hinterlands,
together with those introduced by scores of conquering tribes, now
slowly supplanted the weights and measures of Rome. After the turn
of the millennium rapid metrological growth and proliferation set
in and would gather speed during the Later Middle Ages due to many
factors, the most important being economic development, commercial
competition, demographic growth, increased urbanism, taxation
manipulations, transportation refinements, technological progress,
territorial expansion, and the continuous impact of custom and
tradition.

I shall explore and analyze the dominant characteristics of
these medieval weights and measures by concentrating chiefly on
those employed in the British Isles and France.

Nils Sahlgren

Fil. Dr. / Docent, Stockholm University, STOCKHOLM.

The Secrets of the Old Volume Measures.

In my researches into old <u>cylindrical measures</u> I have proved that to a great extent they were determined on the basis of <u>length</u> measures (inch, foot, ell). The calculations originated from the basic elements of <u>ancient</u> geometry: the <u>straight line</u> and the <u>circle</u>. And they were surrounded by great secrets.

The above mentioned vessels were not only built on the basis of length measures but very often according to geometrical laws, unknown to people today. I have named the <u>three</u> most interesting Models: 1) The Egyptian Triangle Model, 2) The Twelve-Mouth Series, 3) The Skaelskør Model. The names refer to original sources.

1) In Sweden and Northern Germany old cylindrical vessels often are constructed with an "inscribed" so-called <u>Egyptian triangle</u>, with sides = 3, 4 and 5. If such a vessel has height 3 and diameter 4 you get a certain volume. But if you take height 4 and diameter 3 you get a volume that is 3/4 of the first mentioned. Then this vessel has simply been divided into <u>two</u> parts, one half double as big as the other. And than you have got a very useful series of measures with volumes representing 4 : (3) : 2 : 1.

2) An archbishop in Lund (Scania) has in the beginning of thirteenth century described a cylindrical vessel, named "tolvmynning" (Twelve-Mouth) after its <u>diagonal</u> measure, <u>12</u> inches. The height was then <u>6</u> inches. But the Twelve-Mouth appears to be a <u>part</u> of a stereometric series where the basic volume is a vessel with the <u>same</u> height and diameter. If you take the <u>diagonal</u> of this vessel as <u>diameter</u> in another you will get <u>double</u> the volume. And if the same procedure is repeated - and the <u>height</u> of the vessel is the same - the capacity will increase by <u>one basic volume</u> with every "step". In step 3 you will find the Twelve-Mouth. A great number of bigger "relatives" are to be found in Scanian museums.

3) The Skaelskør Model originates from an extremely remarkable carving in the base of the medieval church of Skaelskør in Denmark. According to tradition - since long called in question - the carving shows an <u>ell</u> and a <u>bushel</u>. If you let the circle of the carving represent the bottom of a cylindrical vessel with 1/3 of the ell as height you will find that the <u>whole ell</u> will fit exactly as <u>diagonal</u> in the vessel. But this is not enough: If you let a diametrical section of the vessel be a part of the <u>square</u> drawn around the circle - and let the section "rest" on the middleline of the square - you will find that the upper line of the vessel-section forms a side of square <u>inscribed</u> in the circle. And this square side is <u>2/3</u> of the ell! Further: If you construct a cylindrical vessel round the last mentioned square you get the <u>same volume</u> as I have described above!

The Skaelskør Model was used in Scandinavia, in Northern Germany and - in <u>England</u>. And from there the USA got its measures!

János Farkas

Professor of Sociology, Technical University, Budapest

MAKROSOCIETAL FRAMEWORK OF SCIENCE AND TECHNOLOGY
IN HUNGARY

After World War II. the "big step forward" in the structure of sciences was not organically fit into socio-economic development and has created an "overgrown" science which has proved to be dysfunctional in many respects in the course of later socio-economic development. Thus, a scientific basis and potencial have been created which have grown extensively yet in terms of efficiency we see a sharp decline. On the one hand, in terms of certain scientometric indicators it has become more developed than socio-economic basis on which it has emerged, on the other, in terms of its efficiency, it still lagged behind West Europeans science in spite of good specific indicators.

The conclusion may be drawn that even the copying of the most developed countries' science policy does not yield desirable results. If a country's science is not organically based on the general level of socio-economic development, then forced growth results in much more dysfunctions than in gains. Therefore, more modest and alternative science policies suiting given development levels result in higher officiency than institutional patterns imported and implemented forcibly.

Shigeru NAKAYAMA

University of Tokyo

The Three Stage Development of Knowledge and the Media

John Ziman in his book *Public Knowledge* drew attention to the obvious fact that science is knowledge which only gains significance once it has been published and made available to the public. In contrast to public knowledge we can assume the existence of private knowledge. This term may refer to a number of situations.
The mental satisfaction and philosophical wisdom of life gained through spiritual enlightenment and experience which we keep to ourselves is one example. Knowledge concerning cultural refinement is another. On a different level information related to privacy and the affairs of relatives that is not of public interest also falls within the bounds of private knowledge.
Knowledge including science is usually public knowledge, and by virtue of that fact only attains significance and the ability to be transmitted after it has been published. The form and medium of publication is called the media. Though contents and type of knowledge are the main considerations in selecting a suitable media, we cannot overlook the fact that the media itself does determine knowledge itself. This paper aims to trace the history of the relations between these two aspects of the media.
The term New Media has recently become popular and there are many rumours about its social effects.
Though it is certain that the New Media will influence the transmission of academic information we cannot forsee in which direction it will go. It might even to our surprise change the whole structure of science. Let us consider this possibility by reviewing the past history of the media.

	media	kind	quantity
1.	manuscript	small	small
2.	printing	small	large
3.	new media	large	small